Computational Fluid and Solid Mechanics

Series Editor

K.J. Bathe
Massachusetts Institute of Technology, Cambridge, MA, USA

For other titles published in this series, go to
http://www.springer.com/series/4449

Vladimir Buljak

Inverse Analyses with Model Reduction

Proper Orthogonal Decomposition in Structural Mechanics

Springer

Vladimir Buljak
Politecnico di Milano
Dipartimento di Ingegneria
Strutturale
Piazza Leonardo da Vinci 32
20133 Milano
Italy
buljak@stru.polimi.it

ISSN 1860-482X e-ISSN 1860-4838
ISBN 978-3-642-22702-8 e-ISBN 978-3-642-22703-5
DOI 10.1007/978-3-642-22703-5
Springer Heidelberg Dordrecht London New York

Library of Congress Control Number: 2011941196

Printed on acid-free paper

Springer is part of Springer Science+Business Media (www.springer.com)

Foreword

The methodology of inverse analysis, the origins of which may be regarded as remote and deeply rooted in the history of structural mechanics, has in relatively recent times, emerged as a modern and fast growing area of engineering sciences. In several technological fields, evident are importance and usefulness of reliable transition from experimental data on systems, structures "in primus", to quantitative assessments of crucial properties and possible damages in those systems. Such a kind of assessment means accurate estimations of parameters hidden in mathematical models at present available to simulate and predict the system behavior in service.

The achievement of such ambitious task clearly requires a synergistic convergence of diverse scientific fields: mathematics, usually with their traditional concepts and solution methods (e.g., ill-posedness of problems; minimization/maximization of nonconvex functions), experimental developments in terms of suitable devices, computational techniques and relevant tools.

The growth of computers according to "Moore's law" in the last four-five decades has been, and still is, essential also for the expansion of the inverse analysis developments and applications. Such a link was evidenced in 1986 by Richard Feynman through his celebrated warning on computers ("garbage in, garbage out") after the disaster of the Challenger.

The role of mathematics (specifically, mathematical programming and soft-computing methods, stochastic approaches including Kalman filters, et alia) in inverse analyses is widely important and turns out to be consistent with authoritative recommendations in the engineering history (e.g.: Leonardo da Vinci: "no human investigation is true science if not based on mathematical demonstrations"; Eduardo Torroja: "in the art of building without a mathematical background the designer has no success").

Experiments and measurements, suitably selected by sensitivity assessments, clearly represent the basis of inverse analyses, in full agreement with memorable warnings by great engineers (let us remember Leonardo da Vinci again: "our evaluations may fail, experiments do not"; and Eugène Freyssinet: "when experiment and computation disagree, always computation is wrong"; and also Dan Drucker: "design to design structures is still in large measure based on experience and tests").

The present trends towards synergy of experiments, mathematics and computations, turn out to concern more and more frequently research, engineering practice and scientific education. Popular is becoming at present the following statement by John von Neumann: "science mainly makes models; the justification of such construct is solely that it is expected to work".

This author provides a remarkable contribution to inverse analysis applications in the field of structural mechanics and engineering problems. Peculiar features of this book are the attention paid to the computer implementation of timely procedures for parameter identification and the availability in various chapters of several routines related to popular software like MATLAB and to a widely employed commercial finite element code.

Readers are likely to appreciate detailed treatment of selected inverse problems in structural mechanics and accurate description and employment of representative innovative procedures, practically useful in terms of operative economy (particularly "proper orthogonal decomposition" and "radial basis functions" interpolation). It is desirable that, in future editions of this book, treatments of inverse problems by the same criteria be devoted also to supplementary meaningful subjects and related issues, such as stochastic approaches and relevant computational procedures: these issues are worth of further research efforts and broader applications.

To the study of inverse analysis methodology and to its applications in real life problems, this young author has successfully dedicated his doctoral thesis and his first two years of "post-doc" research and teaching activities in a university environment.

This Foreword ends with cordial, warm wishes to the author for success in his scientific research and teaching career and in future amplifications of this book, dedicated to the growing attractive and productive area of inverse analysis in engineering and technologies.

Giulio Maier

Preface

The goal of this book is to present a modern approach to inverse analyses (IA) that combines traditional framework with numerical techniques used for model reduction. In the main focus are parameter characterization problems in structural mechanics, although most of the material is applicable with slight modifications also to other scientific and engineering fields. The book is intended for engineers and scientist who would like to learn, up to the very details, how to bring together all the necessary pieces into working programs that will solve given inverse problem.

Since the main emphasis was on the implementation, selected algorithms are described into the details required for their implementation, and for all of them practical codes within MATLAB programming language are given with full listings. The codes are written in general way, so it shouldn't be difficult to translate them into any other programming language.

An inverse analyses procedure puts together experimental mechanics, numerical modeling and mathematical programming. For a successful IA procedure one needs to tackle all of these problems. In the structural context discussed in this book, with a traditional approach, simulations of the experiments are done by finite element modeling (FEM), and most frequently commercial codes are used for this purpose. In the problems, tackled within the book, that used this approach to IA a commercial code ABAQUS was selected, while the routines that are written to automatically modify FE models and run the simulations are presented and discussed.

As far as mathematical programming is concerned, given the objective of the book, the most popular optimization algorithms are selected and described up to the details of their successful implementation, while detailed theoretical background descriptions were omitted. Nevertheless, an attempt was made to guide interested readers for useful further readings on the given topics. Optimization algorithms are treated in Chap. 2, and the material in this chapter should serve for the reader to become familiar with all the main concepts of iterative optimization algorithms. The author strongly believes that, after reading this chapter, a careful reader will be able to write his own program that solves numerically an optimization problem by using any of the algorithms discussed in the chapter.

Model reduction technique presented in this book is based on Proper Orthogonal Decomposition (POD) and Radial Basis Functions (RBF). In Chap. 3 it is explained up to very details how these two mathematical techniques are combined into a powerful computing tool that can have an accurate computation of system responses in a computing times shorter by few orders of magnitude with respect to traditional numerical modeling techniques used in structural context (e.g. FE modeling). The construction of proper orthogonal basis in the discreet theory approach was discussed in details, and three different derivations are presented and illustrated with simple examples. The objective of this chapter was to connect the ideas behind the POD theory to the present context and to show how the basic principles developed in different fields can be successfully used also in structural mechanics, and by author's opinion in many other computational problems.

The other mathematical tool used in this reduced basis model, namely RBF interpolation is also described in detailed manner and covered by numerical examples that should serve for a better understanding. Finally, in the last part of Chap. 3, it was demonstrated how the two techniques can be combined into a reduced model used for the computation of system responses in structural mechanics context. This chapter is written with the intentions to explain all the concepts on which reduced basis model here presented is built. The author's opinion is that the careful reader should be capable to, by applying the analogy, employ the described model also to other physical phenomena.

In the last two chapters it was demonstrated how all the previous pieces are put together into a fully working inverse analysis procedure. Chap. 4 showed all the necessary steps for building a so-called traditional IA procedure, where FE simulations are used for the prediction of the system responses. Even though the book presented a modern approach to the inverse analyses, where a reduced basis models should be used for the prediction of system responses, by author's opinion also the traditional approach is very important as it anyhow should be used in some stages of the development. Since the main accent of the book was on the implementation, also in this chapter a detailed description of all the necessary programs was discussed and the developed codes are given in full listings. The material presented in this chapter should be enough for the reader to become familiar with all the elements of practical IA procedure. In the chapter two different case studies are considered that should be used as guidelines for any other similar problem. From this chapter readers should learn how to write from the very beginning a fully working inverse analyses procedure in the structural context by coupling MATLAB routines with commercial FEM code ABAQUS.

Finally, in Chap. 5 of the book it was shown how to incorporate the developed reduced basis model into an inverse analysis procedure. With a fast computational tool like the one developed in Chap. 3, inverse analysis becomes fast and robust. This feature was demonstrated in the examples treated in this chapter. It is shown how to build standalone software which, once that it is calibrated for a given experiment, can be further routinely used on a fast and effective way.

Author believes that the selected material presented in this book should be enough to introduce the readers to the problems encountered in the inverse analysis field. The examples treated in the book should help for a better understanding of all the presented concepts. Author hopes that the book will serve also as inspiration for many different applications of this fast growing scientific field.

Vladimir Buljak

Acknowledgments

I would like to take this opportunity to thank Professor Giulio Maier from Structural Department at Politecnico di Milano, for passing an interest of inverse analyses to me during my Ph.D. work and later, during my "post-doc" phase. His charisma, inexhaustible spirit, dedication to the research, enthusiasm, wide knowledge and above all, friendship have been a great help and inspiration for me.

Special thanks go to Professor Ryszard Białecki from Silesian University of Technology, Gliwice, for his time spent with me during long discussions about Proper Orthogonal Decomposition, on a number of occasions on different conferences. I feel deeply honored for having opportunity to meet and discuss with such a great scientist like Professor Białecki.

I owe my gratitude to Dr. Ziemowit Ostrowski, who I had a fortune to meet. His excellent work was a valuable resource of information for me.

I believe that a significant contribution to this book cames from my colleagues and friends: Dr. Fabrizio Cacchione, Dr. Gabriele Della Vecchia, Dr. Riccardo Rossi, Carlo Guerini, Mohammad Reza Mahini, Dr. Tomasz Garbowski and many others. If this book is written in an understandable way, then it is a merit of these people, who were always finding time for discussions about the topics treated in this book, and who were always asking the right questions that served me as guidelines for the better presentation of the material.

Vladimir Buljak

Contents

Chapter 1
Inverse Analysis: Introduction

Numerical simulations have been established as a powerful tool used in practically all fields of engineering and science. A large number of commercial codes is developed to solve the, so-called *direct problems (or forward problems)*, which consist of finding the solution in terms of response fields when a complete set of input data defining uniquely the solution is known. Since these codes require the knowledge of some parameters on which the solution depends, sometime in engineering practice it is required to solve an *inverse problem*, defined as the one where some of the "effects" (responses) are known but not some of the "causes" leading to them, namely parameters on which the system depends. These problems are tackled within, relatively young and still growing scientific branch which in modern literature (e.g. [1–3]) is found under the name of *Inverse Analyses*.

The name "inverse" comes from the assumption that the logical path in reasoning (or solving the problems) goes from the causes to the effects. Therefore, solving the problem in opposite direction can be identified as inverse (Fig. 1.1).

In the scientific and engineering practice, some observed phenomena are usually represented by models. These models are further used for, hopefully successful, prediction of responses, for a given input parameters. In this context, an inverse problem can be defined as the one in which some model parameters need to be obtained from the given observed data. This definition already anticipates, what in general holds, that to solve an inverse problem is more difficult than to solve the forward one. These difficulties arise from the fact that inverse problems are typically ill-posed, meaning that usually some of the conditions of the uniqueness of the solution do not exist. Certain types of ill-posedness will be discussed in more details later in this chapter.

This book deals with inverse problems that emerge in the structural engineering field. The main idea is to present just the small portion of theory needed to be understood for a successful implementation of a fully working inverse analyses procedure in the present context. This chapter should serve to make the reader familiar with the main concepts and all the necessary parts that one needs to put together in order to solve an inverse problem.

V. Buljak, *Inverse Analyses with Model Reduction*, Computational Fluid and Solid Mechanics, DOI 10.1007/978-3-642-22703-5_1,

Fig. 1.1 Scheme of solving forward/inverse problems

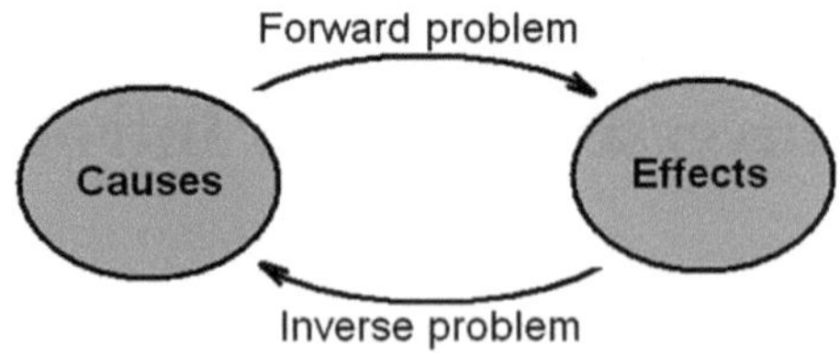

1.1 Inverse Problems in Science and Engineering

We already anticipated that the term *inverse problem* refers to any general situation in which it is required to solve the problem in the opposite direction, starting from "effects" as known and identifying "causes" that lead to them. Inverse problems are inherently connected to direct problems. This connection is established through the model used to simulate a given phenomenon. To solve a direct problem, in scientific or engineering context, means to find analytical or numerical solution for ordinary or partial differential equations that are describing it. On the other hand, an inverse problem is the one in which the solution is known and the objective is to determine the complete forward problem for which that solution is possible.

In order to solve the direct problem there should be known a minimum set of information describing it, referred to as *the condition of uniqueness.* Therefore, when we talk about the solution of differential equations, the uniqueness condition should include the following information:

- The description of geometry under consideration
- Boundary conditions
- Initial conditions
- Properties of all involved materials (i.e. constants entering into differential equations)

Inverse problem is the one in which *some* of these data are missing and they should be identified from the known solution of the problem. Already at this stage some of the problems connected with the inverse analyses can be foreseen. For example it may be possible to face the situation where not only just one set of missing parameters makes the solution possible. It practically means that the inverse problems may not satisfy the uniqueness condition.

This can be illustrated on the following example. Let us consider a simple case in which the model is represented with an analytical equation given by $y = ax^2 + bx + c$. Let the forward solution be the one that uses the value of x to compute y. It is obvious that when the problem is solved as forward, to one value of x corresponds only one value of y (Fig. 1.2). On the other hand the associated inverse problem is not unique, since the solution y_I can be produced by a given model for two different values of x. This lack of uniqueness is common for some inverse problems and as such represents a significant drawback. Further in the book it will be shown how these types of problems are tackled within the inverse analyses in the present context.

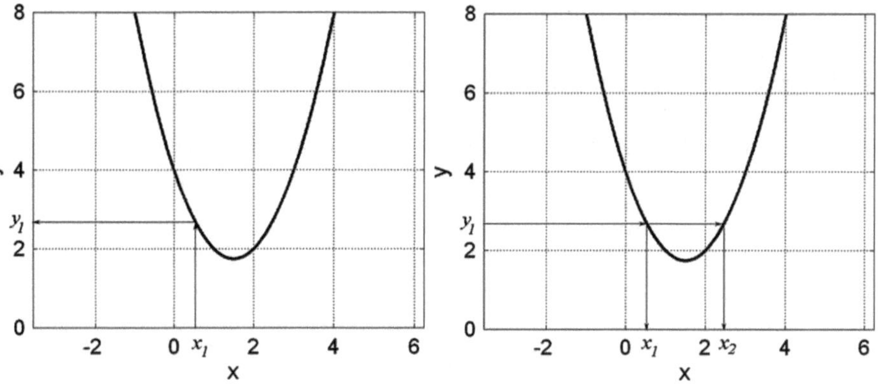

Fig. 1.2 Uniqueness of the solution: existing for the direct problem (*left*) and missing for the associated inverse problem (*right*)

Inverse analyses presented in this book deal with the numerical models. What makes this group of problems particularly interesting and challenging from the engineering and scientific point of view is that they represent a synergic combination between experimental mechanics, numerical simulations and mathematical programming. This combination may be anticipated from the above definition of inverse problems. As mentioned before, the inverse analysis procedure is in general designed in order to assess some of the missing information about the system. The system of interest needs to be modeled by the use of the appropriate technique (in the context presented in this book usually numerical). This model depends on some information (e.g. geometry, boundary condition, parameters, etc.) some of which are known, others should be identified from the known solution. The solution is known as a result of the previously preformed experiment. Finally, a discrepancy function is constructed that quantifies the difference between measured quantities and their computed counter-parts, which is subsequently minimized by the use of optimization algorithms coming from mathematical programming.

1.2 Classification of Inverse Problems and Their Application

There are many ways to classify inverse problems. Going back to their definition, as the problems in which the goal is to identify some missing information, a logical classification could follow based on the type of the missing data. Therefore we can distinguish:

- **Backward problems:** represent problems where the initial conditions are unknown;
- **Boundary inverse problems:** in which the goal is to determine boundary conditions;

- **Shape design:** problems where the shape and the size of the domain need to be determined;
- **Force determination:** problems in which the external action, i.e. forces are unknown;
- **Parameter identification problems:** where constants that are entering into governing equation, primarily into the constitutive models, are to be found.

There are many successful applications of inverse analyses of a different kind. The first two groups of inverse problems cited here are very frequently applied in thermo problems. Using the inverse analyses procedure it is possible to determine both temperatures and heat fluxes on inaccessible surfaces by performing measurements in more appropriate zones (see e.g. [4, 5]). Within Computational Fluid Mechanics (CFD) there are a lot of examples of aerodynamic shape optimization by the use of inverse analyses approach (see e.g. [6, 7]). Force identification problems, are another example of successful application of inverse analyses theory in order to assess, for instance unknown impact or contact forces (e.g. [8]) usually by the measurements of displacements.

In general, inverse analyses are applied in all those fields of science and engineering where the quantities of interest are not directly measurable, and they are therefore assessed through other measured values. Very well established examples of this type can be found in geotechnical engineering considering measurements based on electrical resistivity tomography. This technique provides the measure of conductivity of specimen that is further, through inverse analysis procedure, related to various parameters of soils such as porosity, degree of saturation and hydraulic conductivity (see e.g. [9, 10]).

The last example leads us to, probably the most dispersed group of inverse analyses problems, namely the parameter identification problems. This book will give a particular focus to the parameter identification problems, even though most of the material presented here is applicable also to other groups of inverse problems.

In the last years a big effort was made in the direction of developing numerical models that can reproduce the mechanical behavior of different materials and structures. Nowadays it is possible to use rather complicated material models that are usually already implemented in the commercial codes. These models are capable to capture different physical phenomena (e.g. plasticity, viscoplasticity, damage, fracturing) some times even on different scales. Assuming that they can correctly describe the behavior of the real phenomenon, their accuracy still depends very much on correct identification of the constants (parameters) entering into the governing equation. Sometimes, these parameters are not directly measurable in the laboratories, and so their quantification becomes an important issue. For some simple cases the transition between measured responses in the experiment and required model parameter values can be established relatively easy (like for example construction of a stress–strain curve from measured force and elongation in the uniaxial tensile test). For more complicated material models the link between experiments and computational models is not that trivial. This gap between

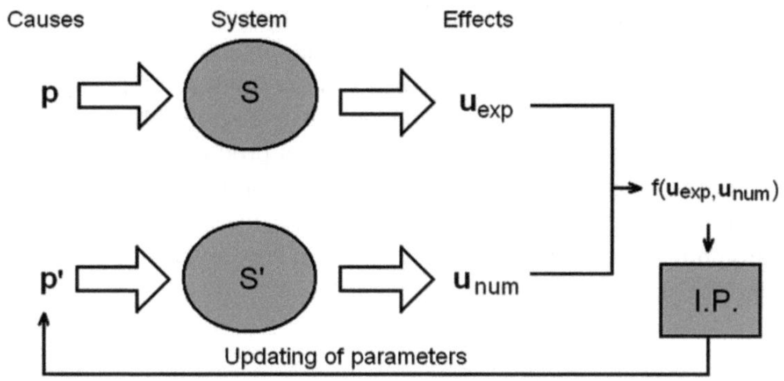

Fig. 1.3 Schematic representation of the inverse analysis procedure

measured material and structural responses and required constitutive model parameters by the computational models is bridged by the inverse analyses theory.

Applying the general concept previously discussed to the parameter identification problems let us denote by S (Fig. 1.3) a certain system, that actually represents an experiment performed on some structure or material specimen. This is the first phase of the inverse analysis procedure that should provide some meaningful measurable quantities. The same process is further modeled (traditionally in the structural context by finite element method) in order to produce a numerical counter part of the system, denoted by S'.

If a perturbation is applied on a real system S (i.e. the system is subjected to a certain load) it will react by giving a response which may be represented by a number of measurable quantities collected in vector u_{exp}. For structural mechanics problems, these can be forces, displacements, crack openings etc. On the other hand, the numerical model is designed to compute the response of the system subjected to the same perturbation, for any given model parameter vector p', collecting the sought parameters within inverse analysis procedure. The response of numerical model of the system represents the calculated values of exactly the same quantities as those measured in the experiment, which are stored in vector u_{num}. The results from the experiment are put together with their computed counterpart in order to form a discrepancy function that should quantify the difference between the two responses (Eq. 1.1)

$$f = \left\| \mathbf{u}_{\text{exp}} - \mathbf{u}_{\text{num}} \right\| \tag{1.1}$$

This function is further minimized with respect to sought parameters. For the minimization a well established algorithms from mathematical programming are usually used, that within some iterations are capable to find parameter values that are minimizing the objective function. In other words the solution of the inverse problem should result with the parameter values for which the adopted numerical model is giving the closest results to those that were measured in the experiment. If the numerical model is done very well, and if the experimental measurements are performed with precise tool with a low level of noise, then at the end of this minimization procedure the numerical result should match exactly the experimental one. In more realistic cases, there will always be a small discrepancy between the two responses, but in any case for well posed inverse problems a result of the minimization should provide parameter values that produce the minimum possible difference for the given measurements and used numerical model.

In order to better understand the main problems one is facing within parameter identification problems let us give a closer look on how an inverse analysis problem is set within the scope of structural mechanics.

1.3 Parameter Identification Problems in Structural Analyses: Setting Up the Problem

Parameter identification problems in material and structural mechanics are used usually in order to characterize the material properties. The approach of inverse analyses gives the possibility to keep the experiment relatively complicated making possible to capture more complex material models described with large number of parameters.

A popular experiment that is used to characterize materials is an instrumented indentation test that originates in the traditional hardness test. It represents the process in which the tip of indenter, usually conical, spherical or pyramidal, is forced into the surface of testing specimen in order to leave a permanent deformation. Unlike the traditional hardness test, in the instrumented one, there is a constant monitoring of applied force and obtained penetration, and so the result of the test is a so-called indentation curve like the one visualized in Fig. 1.4.

The information of material response taken from the indentation curve proved to be rather useful to characterize both elastic and plastic properties. Several authors (see e.g. [11, 12]) have proposed some semi-empirical approaches to correlate the data provided by indentation curves with the Young modulus and yield strength of the material. Using this test within inverse analyses framework turned out to be very much useful making possible to characterize more complicated material models.

As the first improvement with respect to semi-empirical formulae, an inverse analyses procedure can characterize more accurately material parameters just from the indentation curve itself. Since the indentation test can be simulated, with nowadays available commercial FE software, rather accurately, it gives a possibility to compare computed indentation curve with the measured one, without a need to measure contact area and therefore introduce additional potential error of experimental measurements (like in the semi-empirical approach proposed by Oliver and Pharr [11]). A number of successful inverse analyses procedures based on instrumented indentation can be found in the literature: application in characterization of functionally graded materials [13], characterization of thin films [14] assessment of quasi-brittle fracture properties [15], and so on.

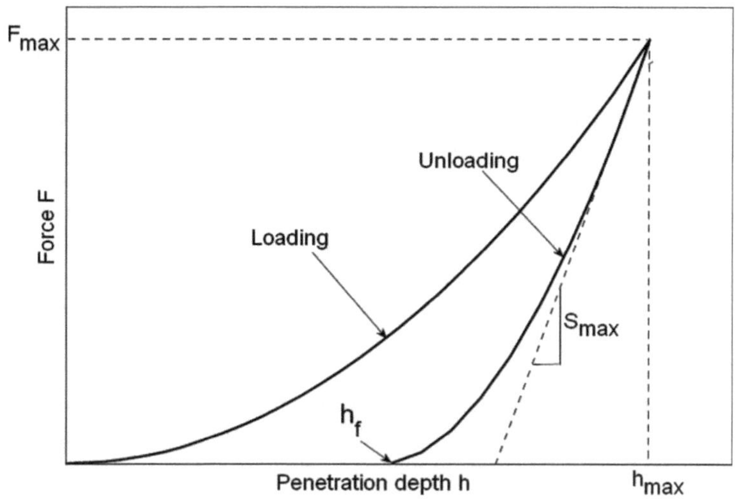

Fig. 1.4 Indentation curve resulting from an instrumented indentation test

Another interesting field of the application of parameter identification procedures is in the assessment of damages in structures that are in service. Since the main concept of inverse analyses is to combine simulation with the experiment, it is evident that the experiment can be rather complicated and performed even in situ directly on the structure. One of the typical examples of this is the assessment of possible deterioration of existing concrete dams due to alkali silica reaction or, in general, due to extreme loadings such as earthquakes and floods. For this purpose a so-called "flat-jack" test is used that consists of performing cuts in the structure (which, considering the size of the dam, are considered as practically non-destructive test); in created slots a flat-jack is inserted that is further pressurized and the resulting displacements are measured. Combining the experiment with numerical modeling of the test a practical inverse analyses procedure can be designed in order to assess the elastic properties of the concrete (see e.g. [16, 17]).

From previous discussion it is obvious that the very first phase in setting the inverse analyses problem for the parameter characterization is to choose the experiment. The experiment should be selected to be as simple as possible but at the same time it needs to activate all the parameters that needs to be assessed. This part is verified usually in a numerical simulation of the experiment.

The parameter identification problems in the context of structural analyses are starting form the assumption that the material model is known and they should result in a calibration of the constants entering into it (e.g. assessment of Young modulus, yield limit, exponent of hardening, etc.). Using the numerical simulation with the adopted material model, it is possible to perform an optimization of experiment in order to find the most appropriate perturbation (i.e. loading pattern of the specimen or structure) and to select those measurements that are the most sensitive to changes of sought parameters.

After the configuration of experiment is decided, the numerical phase is taken a step further in order to perform a so-called sensitivity analyses. Sensitivity analysis is usually carried out over some range of parameter values (the range is adopted based on the expected values of sought parameters for a given problem), and it should result with quantitative information (namely the derivatives of measurable quantities with respect to each of south parameters) about how sensitive measurements are to the perturbations on each of the sought parameters.

Once that the previous numerical study is performed, resulting in the experimental setup from which the measurable quantities with satisfying sensitivity to sought parameters are selected, a subsequent phase proceeds, in which the discrepancy function is formed that represents some norm of a difference between experimental and computed results. It should be mentioned that it is very important to include the uncertainty of the measurements into the objective function. In the form given by Eq. 1.1 if we denote by **R** residual vector that represents the difference between the measured quantities and computed ones, namely

$$\mathbf{R} = \mathbf{u}_{EXP} - \mathbf{u}_{COM}(\mathbf{p}) \tag{1.2}$$

then the objective function (to be minimized) denoted by ω should have the following form

$$\omega(\mathbf{p}) = \mathbf{R}^T \cdot \mathbf{M} \cdot \mathbf{R} \tag{1.3}$$

where **M** is a weight matrix that should somehow take into account uncertainties of the measurements. For example, matrix **M** can be a diagonal matrix assigning to each component of the residual vector **R** a weight inversely proportional to the corresponding measurement scattering. Another possibility could be to represent matrix **M** as the inverse of the (symmetric and positive definite) covariance matrix (diagonal in case of uncorrelated quantities) when each experimental data is provided together with a standard deviation, expressing the uncertainty typical of the measurement device.

To illustrate the importance of weighting the experimental data in the presence of numerical noise, let us consider the following example. Let us assume that the results of some experimental measurements of a certain state variable (e.g. temperature, velocity, etc.) over a space coordinate has the form like the one presented in Fig. 1.5. Let us assume that we would like to predict the distribution of this quantity with an analytical model given by

$$q = p_1 \cdot \sin(p_2 \cdot x) \tag{1.4}$$

The analytical model is governed by two parameters: p_1 and p_2 while x is a spatial coordinate, and q is a quantity of interest. Now we would like to calibrate this model by solving an inverse problem in order to find values of parameters that minimize the difference between experimental data visualized in Fig. 1.5 and those computed by Eq. 1.4. The inverse analyses procedure gives the following values of

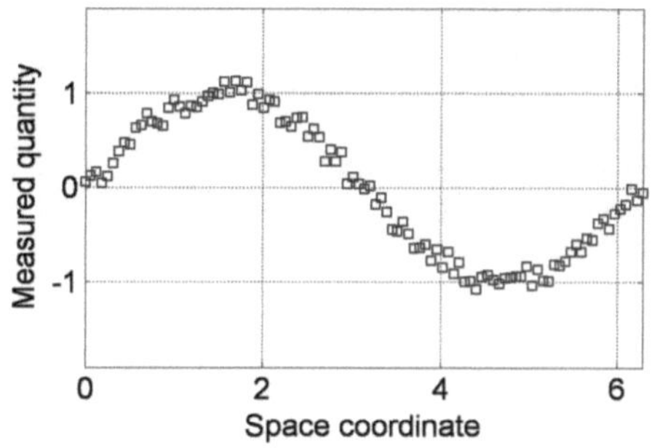

Fig. 1.5 Example of experimental measurements of one quantity in one-dimensional space

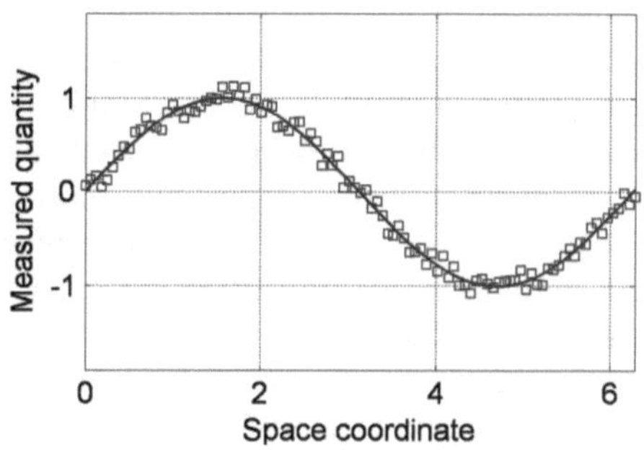

Fig. 1.6 Experimental data and analytical curve that minimizes the objective function

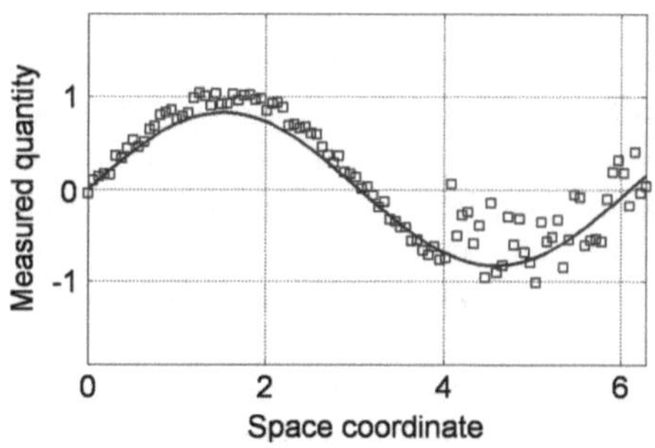

Fig. 1.7 Experimental data with different noise level in different zones, and analytical curve that minimizes objective function without weighting coefficients

parameters that minimize the objective function: $p_1 = 1.013$ and $p_2 = 1.003$ (Fig. 1.6).

Let us assume now that we carry out the same experiment, but this time in the zone where space coordinate is larger than 4 the measurements are performed with larger mistake. Using the objective function that doesn't weight differently the data (say the one defined by Eq. 1.4 with **M** being identity matrix), the procedure converges to different set of parameters, namely $p_1 = 0.828$ and $p_2 = 1.037$ (Fig. 1.7).

As it can be noticed from the figure, the inverse analyses procedure converged to analytical curve that tries to fit both the zone with more accurate measurements and the one with the larger noise. If both of these zones are of equal importance (i.e. no weighting is introduced) resulting parameters that minimize the function are assessed

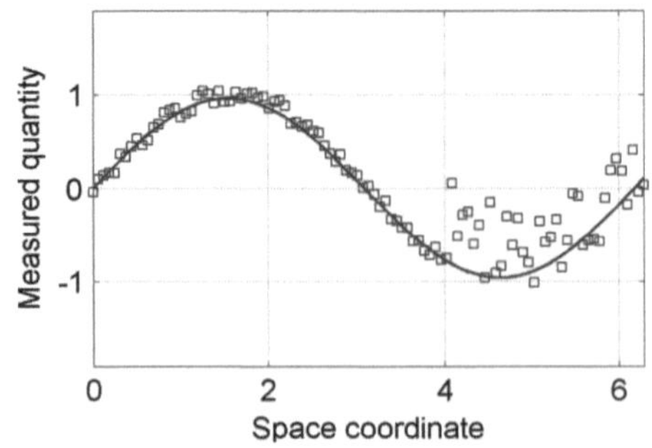

Fig. 1.8 Experimental data with different noise level in different zones, and analytical curve that minimizes objective function with weighting coefficients associated to experimental scattering

with larger error. If we solve now the same problem but this time giving different weighting coefficients to the measurements in the zone $x > 4$ (say three times less weight due to approximately three times larger scattering in that zone) the procedure converges to the following parameter values: $p_1 = 0.960$ and $p_2 = 1.016$ (Fig. 1.8).

Comparing the results it can be noticed that when the experimental data are not weighted, the assessment procedure results in larger errors in estimates, since the function minimizes the difference in an average, least square sense, which can lead to relatively large errors in sought parameters.

Taking into account also the measurement uncertainties is an important issue in all those situations in which the experimental data are not measured with the same accuracy in all zones. This is frequently encountered in the inverse analyses based on instrumented indentation, when the residual imprint is used as experimental data, which is mapped after the removal of the indenter. These measurements are affected with different errors in diverse zones, usually due to optical interferences or other obstacles. In such cases, defining the objective function as a simple difference between measured and computed quantities could result in significant errors on parameter estimates.

Once that the discrepancy function is formulated (namely the measurable quantities are selected and appropriate weighting coefficients are employed) in a subsequent phase, the goal is to minimize this function with respect to sought parameters. In the problems considered in this book the optimization algorithms used for this purpose are numerical, and are not providing the solution in closed form. These algorithms are traditionally iterative and they seek the minimum of the objective function by a successive approaching within a sequence of calculations.

The minimization problems dealt within the context discussed here are always constrained, since the parameters are representing some physical quantities (i.e. Young modulus, yield limit, etc.) and therefore cannot take any arbitrary values. What makes the optimization problem even more complicated is that the objective function, in most of the cases is not convex, and sometimes even not continuous.

There are many different algorithms that one may employ to solve the constrained minimization problem. The choice of appropriate algorithm for the given problem may have a crucial effect on the overall effectiveness of the whole inverse analyses procedure. Without entering into a detailed survey of all possible choices with their advantages and drawbacks at this stage let us just point out some more important features that need to be verified.

Among many possible classifications of optimization algorithms let us for a moment focus our attention to the classification based on the information on the objective function required by the algorithm. Taking this criterion into account all the algorithms can be classified in the following groups: **zero order algorithms**, which are involving only calculations of the objective functions (e.g. direct search by Nelder-Mead algorithm, Genetic Algorithms etc.); **first order algorithms** that need first derivatives of the objective function (e.g. steepest descent method, trust region method) and **second order alogrithms** requiring information on the Hessian, namely second derivatives of the objective function (e.g. Newton method).

Considering that the function to be minimized in our case is not analytically defined, it means that the computation of the derivatives is performed numerically, usually by finite differences. It implies that to compute the first derivatives of the objective function that depends on two parameters, it is required to perform three computations of system responses (namely three simulations): one for the reference values of the parameters, and remaining two for the separate perturbations of the two parameters. The fact that the derivatives are computed numerically already penalizes the second order methods, since using finite differences to computed full Hessian matrix for the two parameter case will involve six computations (with respect to three needed for first derivatives). Due to the increased computing times connected with second derivatives usually the second order methods are not used, since their potentially better behavior is penalized by the increased computing times.

In general, first order methods are less costly computationally than zero order methods (namely they involve fewer simulations). These algorithms start from some initial guess for the parameters, and then in each iteration they compute the first derivatives of the objective function. Based on these computations an evaluation of the parameter values for the next iteration is performed. The general scheme for first order algorithms is presented in Fig. 1.9.

Relaying on first derivatives only, these methods are incapable of distinguishing between the local and global minima. In other words the first order methods will converge to any mathematical minimum of the function (namely the one for which the first derivatives are equal to zero). The fact that the minimum is not a global one can be easily verified since the computed response for the converged parameters will differ from the expected one. In such cases the optimization procedure should be repeated starting from different initialization (i.e. different guess parameters). The fact that algorithm converged to the local minimum doesn't mean that the parameter estimate procedure doesn't work. Since it can be easily verified it practically represents only a loss of computational time. However, if the objective function for the given problem turns out to have a large number of local minima, then the approach with the use of first order algorithms will result as ineffective since the solution will depend on the initialization point. Even though the local minima can be verified as such, and therefore it may be confirmed that the set of parameters is not the solution we were looking for, still the presence of large

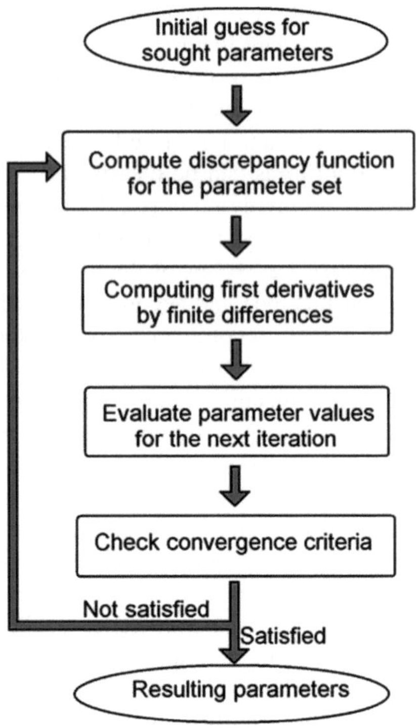

Fig. 1.9 General scheme of first order optimization algorithms

number of local minima makes the procedure not very effective since it may require a large number of initializations until it finally finds the global minimum.

If it turns out that the objective function to be minimized has a large number of local minima, then it is more appropriate to use a zero order optimization algorithms, like for example Genetic Algorithms (GA). These algorithms belong to a so-called soft computing group and will be discussed in more details in the following Chapter. At the moment let us just anticipate that GA, in general, avoid stacking in the local minima and are therefore much more effective if the objective function has a multiple local minima. This nice feature however, comes for the price of increased computing times and therefore the use of GA is justifiable only in the optimization problems with a large number of local minima. In other situations it is more convenient to use first or second order algorithms.

More serious problem which can be encountered within the inverse analysis procedure is the presence of more global minima in terms of two or more parameter sets that produce the same response of the system. In the shape optimization problems this doesn't represent a serious malfunction of the procedure since the goal is to find the shape that optimizes given requirements. An example of this can be a shape optimization problem with the goal of overall weight minimization under constrain of prescribed admissible stresses. If the objective function turns out to have two or more global minima in this case it means that the same weight can be

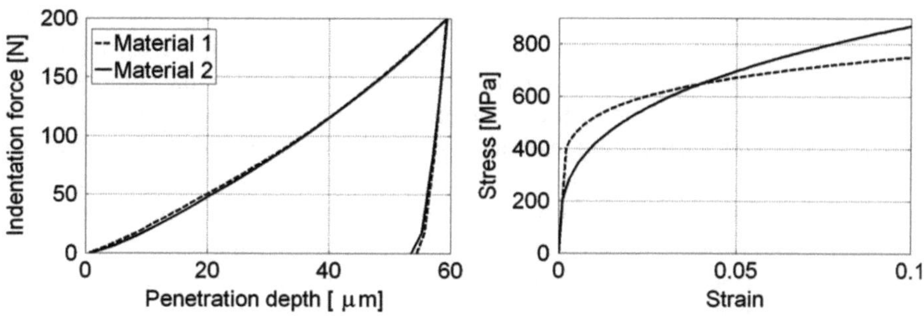

Fig. 1.10 The same response to the indentation test (*left*) of two different materials with different stress–strain curves (*right*)

obtained by different shapes and then it is on the designer to choose the one, so the optimization problem is considered to be solved.

In the material characterization problems the presence of two or more global minima represents a serious drawback of the procedure since it practically means that there are two possible material parameter sets which can produce the same response to the selected experiment. Typical example of this is a response to the indentation test in terms of the indentation curve. Figure 1.10 visualizes the simulation of the indentation test of two different materials with stress–strain curves visualized on the right-hand side of the figure. The left-hand side graph shows that the indentation curves are practically the same. This unpleasant feature of the indentation test is evidenced in the literature (see e.g. [18]) where the authors addressed this problem, and labeled materials as "mystical", since they appear to be undistinguishable.

From the inverse analyses point of view this represents a typical problem of ill-posedness with the discrepancy function having two (or more) global minima. Unlike the case with local minima, that can be solved by changing the optimization algorithm, in the case of more global minima nothing can be improved by changing the optimization algorithm, since, from mathematical point of view both parameter sets minimizes the function and it's not possible to distinguish which one is a solution for the inverse problem. The core of the problem is in the objective function itself, or to be more precise, in the selection of measurable quantities used to construct it. The only way to change it is to take into account additional measurements (or even additional experiment) in order to enrich the experimental information. This usually helps since in general, it is unlikely that two different materials will have the same response in terms of all the measurable field variables, or to two different experiments.

In the context of previously discussed undistinguishable materials in terms of response to the indentation test, Bolzon et.al. in [19] showed that this problem is regularized if also the residual imprint is measured and added to the objective function. Physically this means that even though the two materials (like those in Fig. 1.10) can produce the same indentation curve, there will be a difference in terms of residual imprint geometry. This intervention changed the objective function which became more regular and doesn't have anymore multiple global minima.

The problem of more than one global minimum is produced by the compensation of influences between some of the parameters (like in the example of indentation test between yield stress and exponent of hardening). As such, it cannot be identified in the sensitivity analyses phase. The sensitivity analysis is done in order to verify that the selected measurements are influenced by the changes on the parameters. However, even if they are influenced by the parameter perturbations it may occur that some of the effects can be compensated producing therefore more than one set of parameters that minimize to the same level the objective function.

This suggests that, after all of the previously described steps are performed, a final check of the whole procedure should be performed. This part is done using the so-called pseudo-experimental data.

The only way to confirm that the inverse problem is set well and that it is capable of finding the sough parameters, is by solving it for the case where the solution is known. Therefore we should provide the system response for which the parameters are know. Since for this purpose truly experimental data with known parameters may not be available, it is more convenient to use pseudo-experimental data. Pseudo-experimental data represent computer generated data, resulting from the simulation with known parameters. When such data are used as inputs to the inverse analyses, the solution is known, and if the procedure is set well it should result with the same parameters as those used to generate pseudo-experimental data.

Furthermore, such verification should be performed for various parameter combinations, spread around the zone of interest, to confirm that the problem is well-posed not only just in one zone. It is important to do so since, the previously discussed problems of compensation between parameters may occur for some parameter combinations and not for others. As a result of these checks usually graphs of the type presented in Fig. 1.11 are designed for each of the parameters to evidence the error on estimates.

The result visualized in Fig. 1.11 represents an output of well set inverse problem. It shows that the parameter 1 is identified with relatively uniform error, for 50 different parameter combinations (presented in the abscissa). The error is uniform practically in all of the zones and it doesn't exceed the level of 1% (ordinate).

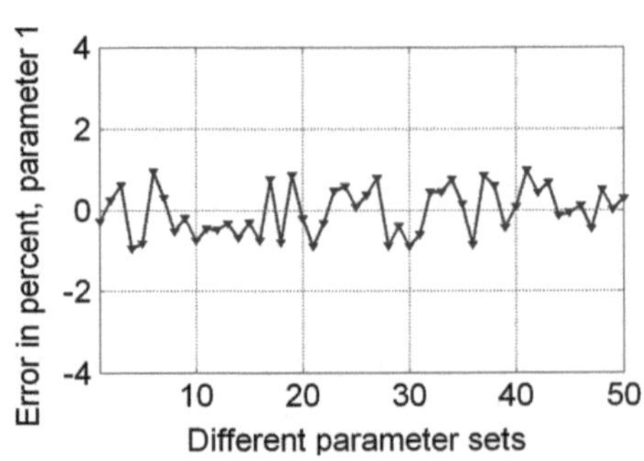

Fig. 1.11 Typical representation resulting from pseudo-experimental testing

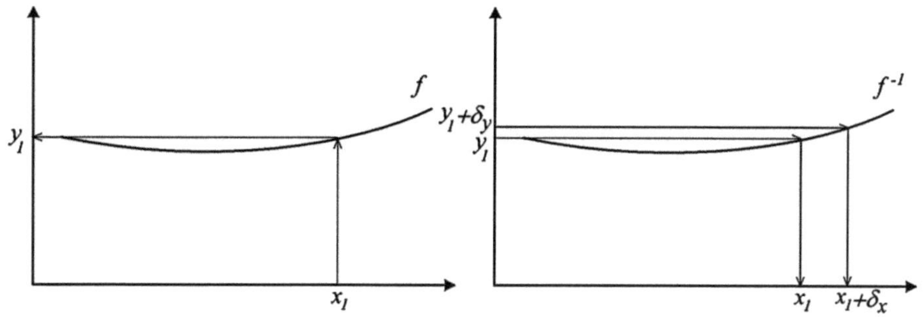

Fig. 1.12 Stabile solution for the direct problem (*left*) and not stable for the associated inverse problem (*right*)

This first pseudo-experimental check was dealing with "clean" numerical data used in the form resulting from the numerical simulations. Since the real experimental data are always subjected to a certain measurement error, it is important to check how the inverse procedure will behave in the presence of noise. Some times inverse problems don't have the same stability of the solution as direct problems (namely to the small perturbation of inputs there is a corresponding small variation of outputs, see Fig. 1.12). Since these perturbations are expected to occur due to errors in measurements it is important to check how the procedure reacts to them. This check should be performed with pseudo-experimental data, by subjecting the inputs, resulting from previously performed simulations, to a random "noise" that should imitate the excepted measurement error. After this intervention, the same study is performed which results in the graphs of the type given in Fig. 1.11, constructed for the different noise levels. From this test, it should be confirmed that the perturbation of the inputs of the certain level should result with more or less the same level or error of the estimated parameters. This study should provide valuable information on the required level of accuracy of the measured quantities.

Another possible resource of error in the identification procedure can come from the numerical simulation. Even when all the quantities are known for given experiment (namely, geometry, loads, constants entering into material models etc.) there will always be an error in modeling resulting in slightly different simulated response from the one measured in the experiment. This modeling error is rather systematic and within inverse analysis procedure can produce the same, systematic, type of error on the estimates.

To illustrate the influence of modeling error let us consider a simple case in which an experiment is performed on a cantilever, of the unknown elastic properties, with the concentrated load at the end. The discrepancy function is formed using, say ten measured displacements along the upper surface. Let us assume that the goal is to design an inverse analysis procedure that should minimize this function in order to assess the Young modulus. If the cantilever is modeled with the coarse FE mesh with a small number of elements over the thickness (see Fig. 1.13), the numerical model will be more rigid than the actual one, resulting

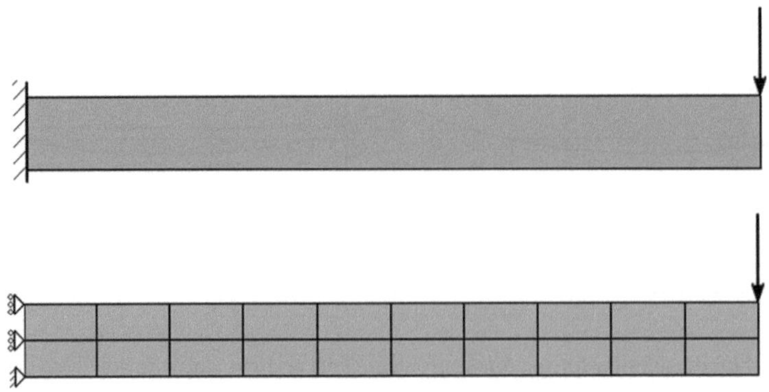

Fig. 1.13 Scheme of cantilever experiment, and numerical model with the coarse mesh

in smaller displacements. Using such model within inverse analyses procedure will produce a systematic error on estimated Young modulus resulting in smaller value than it should be, since the procedure searches the parameter for which the result will match the measured one. An over-stiff behavior is therefore compensated by a softer material. This brings a systematic error to the identification procedure resulting in underestimation of the parameters even when the experimental measurements are performed precisely.

The only way to check how good numerical model is would be to perform the experiment on the specimen with known material parameters and to compare it with the simulation. In the cases when such experimental data are not available, similar check, at least in qualitative terms, can be performed using pseudo-experimental data. It is done by perturbing the simulated data with a certain percentage, but this time with systematic error (i.e. on one side, like increase or decrease computed values resulting from the simulations). Using such data puts in evidence how the modeling error of given magnitude will affect the estimates of the parameters. If they remain within acceptable range it means that no further improvements are necessary on the numerical model. In opposite case it means that there is no significant stability of the given inverse problem and so the accuracy of the numerical model plays an important role. In such circumstance the modeling error should be verified before using the model in the context of inverse analyses.

1.4 Summary

In this Chapter a brief introduction to the inverse problems was given. A general classification and some of the applications in engineering and science are mentioned. A bit more detailed description of the parameter identification problems, as they are the main topic of this book, is then presented.

In the context of structural problems, it was shown which main ingredients need to be put together in order to design a fully operative identification procedure. For this purpose, a sequence of steps and procedures one needs to perform, together with the main possible problems which can be faced, is presented. This sequence can be summarized as follows:

1. The first step of each parameter identification procedure is the selection of the experiment. The experiment should be simple enough in order to be easily executed, but at the same time complicated enough to activate all the sought parameters for which the procedure is designed;
2. In the second step a numerical model of the experiment is build. Traditionally the FEM is used in the structural context. Numerical model should represent a compromise between the required level of accuracy, and the computing time having in mind the repetitive simulations required by the optimization algorithms;
3. Once that a reliable numerical model is constructed, it should be used to perform sensitivity analyses to verify that the selected measurable quantities are sensitive to the variation of parameters. If they turn out to be not sensitive enough they should be changed, or the experiment should be replaced with another one;
4. After the experiment setup is fixed, the objective function is designed that should quantify the discrepancy between the measured quantities and their computed counter-parts. It is important to apply the correct weighting coefficients that should account for measurable uncertainties. Further, an adequate optimization algorithm should be selected for the given problem with primer focus on the possible presence of local minima.
5. Finally, the whole procedure should be tested using pseudo-experimental, computer generated data. This step should put in evidence a possible ill posedness of the inverse problem in terms of the existence of more than one parameter set that minimize to the same extent the objective function (more global minima). If this turns out to be the case the experimental measurements should be enriched, or even an additional experiment should be considered. The pseudo-experimental testing should also be performed using different levels of "noise" (both randomly distributed, to simulate the expected error of measurements, and systematic one, to account for possible modeling errors). The test with noisy data should show how stabile the solution is with respect to the small perturbations of the inputs.

From this chapter it should be clear that one of the crucial parts of any inverse analysis procedure that determines to a large extent the overall success of the whole process is the optimization algorithm. Therefore, subsequent chapter will give a more detailed survey of different optimization algorithms which can be used in the present context.

References

1. Groetsch, C.W.: Inverse Problems: Activities for Undergraduates. The Mathematical Association of America, Washington, D.C. (1999)
2. Liu, G.R., Han, X.: Computational Inverse Techniques in Nondestructive Evaluation. CRC Press, Boca Raton (2003)
3. Tarantola, A.: Inverse Problem Theory and Methods for Model Parameter Estimation. Society for Industrial and Applied Mathematics, Philadelphia (2005)
4. Ostrowski, Z., Bialecki, R.A., Kassab, A.J.: Estimation of constant thermal conductivity by use of proper orthogonal decomposition. Comput. Mech. **37**(1), 52–59 (2005)
5. Yu, G., Wen, P.A., Wang, H.: An inverse method to determine boundary temperature and heat flux for a 2D steady state heat conduction problem. Proceedings of the ASME International Design Engineering Technical Conferences and Computers and Informational in Engineering Conference (IDETC/CIE 2008), New York. Paper number DETC2008-49811, pp. 1087–1093 (2008)
6. Li, W., Krist, S.A., Compabell, R.: Transonic airfoil shape optimization in preliminary design environment. Proceedings of 10th AIAA/ISSMO Multidisciplinary Analysis and Optimization Conference, Albany. Paper number AIAA 2004-4629, pp. 3650–3671 (2004)
7. Sobieczky, H.: Knowledge based aerodynamic optimization. Proceedings of 4th SST CFD Workshop, Tokyo, pp. 1–6 (2006)
8. Wang, B.T., Chiu, C.H.: Determination of unknown impact force acting on a simply supported beam. Mech. Syst. Signal Process. **17**(3), 683–704 (2003)
9. Borsic, A., Comina, C., Foti, S., Lancellotta, R., Musso, G.: Imaging heterogeneities with electrical impedance tomography: laboratory results. Geotechnique **55**(7), 539–547 (2005)
10. Damasceno, V., Fratta, D.: Monitoring chemical diffusion in a porous media using electrical resistivity tomography. ASCE Geotech. Spec. Publ. (GSP) **149**, 174–181 (2006)
11. Oliver, W.C., Pharr, G.M.: An improved techniques for determining hardness elastic modulus using load and displacement sensing indentation experiments. J. Mater. Res. **7**, 176–181 (1992)
12. Jager, A., Lackner, R., Eberhardsteiner, J.: Identification of viscoelastic properties by means of nanoindentation taking the real tip geometry into account. Int. J. Theor. Appl. Mech. AIMETA **42**(3), 293–306 (2007)
13. Nakamura, T., Wang, T., Sampath, S.: Determination of properties of graded materials by inverse analysis and instrumented indentation. Acta Mater. **48**, 4293–4306 (2000)
14. Van Vliet, K.J., Gouldstone, A.: Mechanical properties of thin films quantified via instrumented indentation. Surf. Eng. **17**(2), 140–145 (2001)
15. Maier, G., Bocciarelli, M., Bolzon, G., Fedele, R.: Inverse analyses in fracture mechanics. Int. J. Fract. **138**, 47–73 (2006)
16. Fedele, R., Maier, G.: Flat-jack test and inverse analysis for the identification of stress and elastic properties in concrete dams. Meccanica **42**, 387–402 (2007)
17. Fedele, R., Maier, G., Miller, B.: Identification of elastic stiffness and local stresses in concrete dams by in situ tests and neural networks. Struct. Infrastruct. Eng. **1**(3), 165–180 (2005)
18. Chen, X., Ogasawara, N., Zhao, M., Chiba, N.: On the uniqueness of measuring elastoplastic properties from indentation: the indistinguishable mystical materials. J. Mech. Phys. Solids **55** (8), 1618–1660 (2007)
19. Bolzon, G., Maier, G., Panico, M.: Material model calibration by indentation, imprint mapping and inverse analysis. Int. J. Solids Struct. **41**, 2957–2975 (2004)

Chapter 2
Optimization Algorithms

This chapter will give a survey of some of the most commonly used optimization algorithms within the context of parameter characterization. The idea is not to give any detailed mathematical description of numerical optimization, since, on that topic one can found a decent number of great books (e.g., [1–3]). As the main topic of this book are the inverse analyses in structural engineering context, the goal is to present, to a reasonable extent, mathematical theory behind most commonly used optimization algorithms, so that they can be understood and easily implemented into a practical inverse analysis procedure.

Mathematically speaking, optimization is the minimization or maximization of a function subjected or not to constraints on its variables (parameters). In order to solve any optimization problem numerically, nowadays there is a wide variety of algorithms at our disposal. As we already saw in the previous chapter, these algorithms are starting from some *initial guess* of the parameters, and then they generate sequence of iterates which terminates, when either no more progress can be made, or when it seems that a solution point has been approximated with sufficient accuracy. Within each iteration, certain information on objective function are gathered (i.e. the value of objective function, the values of first derivatives, the values of second derivatives), and based on these computations the new iterate with a lower function value is estimated. It should be mentioned however, that there also exist *non-monotone algorithms* that do not insist on a decrease in the objective function at every step. No matter if the algorithm leads to a monotone decrease of the objective function of not, the main difference between the optimization algorithms is the way on which they pass from one iteration to another. This characteristic of the algorithm determines to a large extent its performance for a given optimization problem.

In the following pages two different strategies for computing next iteration from the previous one will be presented in more details, as they are used most frequently in nowadays available optimization algorithms. The first one is the *line search* strategy in which the algorithm chooses a direction $\mathbf{p}_k$ and then searches along this direction for the lower function value. The second strategy is called the *trust region* in which the information gathered about the objective function is used to construct a model function whose behavior near the current iterate is *trusted* to be similar

V. Buljak, *Inverse Analyses with Model Reduction*, Computational Fluid and Solid Mechanics, DOI 10.1007/978-3-642-22703-5_2,

enough to the actual function. Then the algorithm searches for the minimizer of the model function inside the trust region.

The last part of this chapter is devoted to a brief description of Genetic Algorithms, as they can be very useful in the problems when the objective function has a lot of local minima.

Before describing above mentioned optimization algorithms let us first say couple of words about least squares problems, or the problems in which the objective function has a form of summation of the squares of differences. As it was shown in Chap. 1, the function to be minimized that emerges form the inverse analyses (Eq. 1.1) is of this type.

2.1 Least Squares Problems

In least squares problems the objective function to be minimized has the following form

$$f(x) = \frac{1}{2}\sum_{i=1}^{m} r_j^2(x) \tag{2.1}$$

where r_j are called residuals.
The objective function of the least squares type emerges from many problems, and probably represents the most frequent optimization problem. In any engineering or scientific field where a parameterized models are used to fit the actual data, the function of the type (2.1) is used to measure discrepancy of the computed quantities and those that are measured.

If we compare function (2.1) with (1.1) from the previous chapter we can notice that it is of the same type. In this case, each residual represents the difference between computed quantities and their measured counter-part, and therefore represent the function of sought parameters.

To see why this special form of the objective function usually makes the least squares problems easier to solve than the general ones, let us first collect all the individual components into a residual vector **R**, namely

$$\mathbf{R}(x) = [r_1(x), r_2(x), \ldots, r_m(x)]^T \tag{2.2}$$

Using this notation, the objective function can be written as

$$f(x) = \frac{1}{2}\|\mathbf{R}(x)\|^{L^2} \tag{2.3}$$

The derivatives of the objective function can be expressed in terms of the Jacobian matrix **J**, which, in the case when **x** is n-dimensional vector will be $m \times n$ matrix, namely

$$\mathbf{J}(\mathbf{x}) = \left[\frac{\partial \mathbf{R}}{\partial \mathbf{x}}\right] = \begin{bmatrix} \frac{\partial r_1}{\partial x_1} & \frac{\partial r_1}{\partial x_2} & \cdots & \frac{\partial r_1}{\partial x_n} \\ \frac{\partial r_2}{\partial x_1} & \frac{\partial r_2}{\partial x_2} & \cdots & \frac{\partial r_2}{\partial x_n} \\ \cdots & \cdots & \cdots & \cdots \\ \frac{\partial r_m}{\partial x_1} & \frac{\partial r_m}{\partial x_2} & \cdots & \frac{\partial r_m}{\partial x_n} \end{bmatrix} \tag{2.4}$$

The gradient of the objective function can be written in terms of the Jacobian as

$$\nabla f(\mathbf{x}) = \sum_{i=1}^{m} r_i \frac{\partial r_i}{\partial \mathbf{x}} = \sum_{i=1}^{m} r_i \begin{bmatrix} \frac{\partial r_i}{\partial x_1} \\ \frac{\partial r_i}{\partial x_2} \\ \cdots \\ \frac{\partial r_i}{\partial x_n} \end{bmatrix} = \mathbf{J}^T \cdot \mathbf{R} \tag{2.5}$$

Hessian matrix of second derivatives of the objective function can be written as

$$\nabla^2 f(\mathbf{x}) = \sum_{i=1}^{m} \frac{\partial r_i}{\partial \mathbf{x}} \cdot \left(\frac{\partial r_i}{\partial \mathbf{x}}\right)^T + \sum_{i=1}^{m} r_i \frac{\partial^2 r_i}{\partial \mathbf{x}^2} = \mathbf{J}(\mathbf{x})^T \cdot \mathbf{J}(\mathbf{x}) + \sum_{i=1}^{m} r_i \frac{\partial^2 r_i}{\partial \mathbf{x}^2} \tag{2.6}$$

What can be observed from the Eq. 2.6 is that the part of the second derivatives can be expressed by the Jacobian matrix. It practically means that, once the first derivatives are computed, we can also compute part of the Hessian matrix for the same computational cost. The possibility to compute "for free" the Hessian matrix once the Jacobian is available represents a distinctive feature of least squares problems. Since near the solution the residuals are close to zero, it also means that the contribution of the second part of Hessian matrix is very small. Therefore, it is reasonable to approximate the Hessian matrix with the first part only.

$$\nabla^2 f(\mathbf{x}) \approx \mathbf{J}(\mathbf{x})^T \cdot \mathbf{J}(\mathbf{x}) \tag{2.7}$$

This approximation is adopted in many applications as it provides an evaluation of the Hessian matrix without computing any second derivatives of the objective function.

2.2 Line Search Method

Within the line search method, in each iteration it is required to find the direction, say $\mathbf{p}_k$ and then to decide how far to go along that direction. Therefore, if the current iteration is denoted by vector $\mathbf{x}_k$, the new iteration is given by

$$\mathbf{x}_{k+1} = \mathbf{x}_k + \alpha_k \mathbf{p}_k \tag{2.8}$$

The new iteration is thus uniquely defined by the direction $\mathbf{p}_k$ and by a positive scalar α_k called the step length. The success of a line search method depends on effective choices of both the direction and the step length. Based in the adopter strategy of solving for the two abovementioned quantities, we can distinguish between different lines search algorithms.

2.2.1 Line Search with Steepest Descend Direction

Most of the line search algorithms are implemented so that they have a monotone decrease of objective function. This means that the direction $\mathbf{p}_k$ needs to be a descending direction. Therefore the most logical direction to move along would be the *steepest descent* direction or the one that defines direction $\mathbf{p}_k$ as

$$\mathbf{p}_k = -\frac{\nabla f_k}{\|\nabla f_k\|} \tag{2.9}$$

The advantage of this choice is that it involves the calculation of only first derivatives.

After the direction is fixed, the algorithm needs to compute the step length. In the step length selection we are facing a tradeoff. Obviously we would like to have as good as possible reduction of the objective function, but at the same time we don't want to spend a lot of time in searching for it. The ideal case would be to find a step length as a global minimizer of the following function

$$\phi(\alpha_k) = f(\mathbf{x}_k + \alpha_k \mathbf{p}_k) \tag{2.10}$$

However, it may be computationally too expensive to identify this value. Because of that, most practical strategies that use the line search approach are finding an inexact minimizer of (2.10) for the acceptable computational cost. This minimizer should satisfy some criteria for the function reduction but without a guarantee that it is the global minimizer of the function along that direction. Typical line search algorithms try a sequence of candidate values for the step length and then select the best of them.

In order to see how the step length influences the performance of line search method let us consider a simple example. Let the objective function be given by

$$f(\mathbf{x}) = x_1^2 + x_2^2 \tag{2.11}$$

The function to be minimized is convex and has only one minimum corresponding to coordinates $\mathbf{x} = [0,0]^T$. Figure 2.1 visualizes the function on the domain-2 to 2.

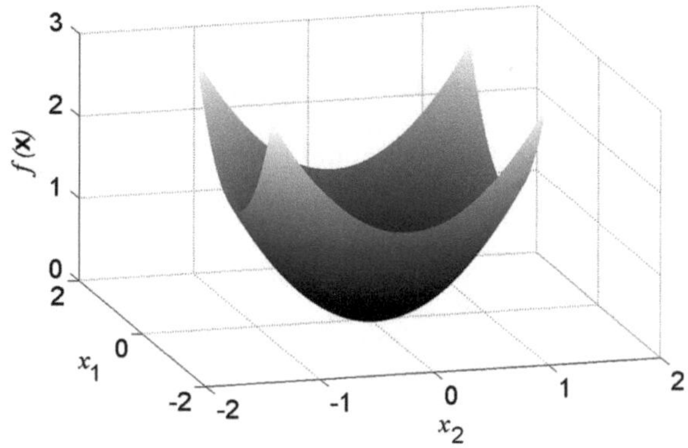

Fig. 2.1 The objective function given by Eq. 2.11 on the domain $x_1, x_2 \in [-2, 2]$

In the following MATLAB code a simple strategy with three trial step lengths is implemented. Within this strategy the three trial steps are: the reference step length, one-quarter of it, and two times reference length. The algorithm starts in the first iteration with some initial step length, taken as a reference one, and for each of the three trial steps it computes the reduction of the function. It further picks up the step with the largest reduction of the function and in the following iteration adopts it as a reference one. This simple strategy allows for continuous elongation (or shortening) of the step length with respect to the starting value, in order to make better use of the selected direction.

Listing given in the following page includes three MATLAB codes. The first one is the main optimization routine, the second one is used to plot the results and the third one is a MATLAB function that computes the value of the objective function for given parameters.

In the main routine the initial step length, together with other options, is chosen by the user at the beginning of the code. The optimization is terminated when the residual is smaller than a given value. The optimization path is recorded in the matrix `itiner` that has the number of rows equal to the number of iterations. The first derivatives are computed my finite differences, even though in this simple case they can be computed analytically. In more general cases, that will be discussed further in the book, the objective function will not be given in analytical form and therefore the derivatives will always be computed by finite differences.

Second MATLAB code serves for the visualization of the optimization results as mentioned previously. One optimization result is given in Fig. 2.2 that shows the convergence after 8 iterations starting from the point given by $\mathbf{x} = [1.1,\ 1.5]^{\mathrm{T}}$, for the initial step length equal to 0.1. It may be observed from the figure that, from the first to the fourth iteration the algorithm constantly enlarges the step length after which it starts to shorten it as it approaches the solution. The selection of the initial step is not influencing very much the performance of the optimization algorithm as long as it is selected within a reasonable range for the given problem. Table 2.1 shows the optimization path for this problem for the two different initial step lengths: $\alpha_{\mathrm{IN}} = 0.1$ and $\alpha_{\mathrm{IN}} = 0.5$. It may be observed that, even though the steps

```
*************************************************************
% Optimization algorithm based on line search method.
% Step length identification by three trial steps

% Setting the options for the optimization
guess=rand(2,1)*2-ones(2,1); % Initial vector of parameters
pert=0.001; % Perturbation for the first derivatives
length=0.1; % Initial step length
resMIN=1e-6; % The value of residual at the termination

% Optimization cycle
res=10;
iter=0;
while res>resMIN
    itiner(iter+1,1:2)=guess';
    e=exfun(guess);
    itiner(iter+1,3)=e;
    iter=iter+1;
    for i=1:size(guess,1)
        guessp=guess;
        guessp(i)=guessp(i)+pert;
        e1=exampl2(guessp);
        grad(i,1)=(e1-e)/pert;
    end
    stpdsc=-grad/norm(grad);
    % Trying different step lengths
    guess1=guess+length*stpdsc;
    guess2=guess+(length/4)*stpdsc;
    guess3=guess+(length*2)*stpdsc;
    e1tr=exfun(guess1);
    e2tr=exfun(guess2);
    e3tr=exfun(guess3);
    best=min([e1tr,e2tr,e3tr]);
    if e1tr==best
        guess=guess1; step=1;
    end
    if e2tr==best
        guess=guess2; step=2; length=length/4;
    end
    if e3tr==best
        guess=guess3; step=3; length=length*2;
    end
    res=guess'*guess; % Computing residual
    itiner(iter+1,1:2)=guess';
    itiner(iter+1,3)=eTR;
end
*************************************************************

*************************************************************
```

```
% Plotting the results of the optimization
N=0;
for i=-1.2:0.01:1.2
    N=N+1;
    M=0;
    for j=-1.2:0.01:1.2
        M=M+1;
        fun(N,M)=exfun([i;j]);
    end
end
i=-1.2:0.01:1.2; j=-1.2:0.01:1.2;
figure(1)
surf(i,j,fun,'LineStyle','none')
grid on
hold on
plot3(itiner(:,1),itiner(:,2),itiner(:,3),'MarkerSize',10,'Mar
ker','o','LineWidth',3,'LineStyle','--','Color',[0 0 0])
view([-1,1,3])
hold off
************************************************************

************************************************************
% Objective function
function e=exfun(x);
e=x(1)^2+x(2)^2;
************************************************************
```

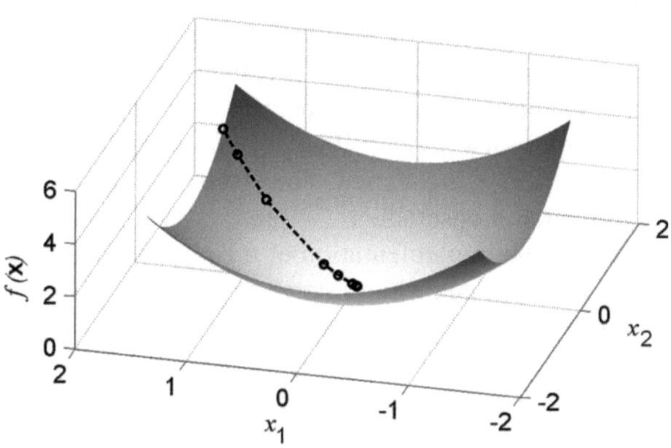

Fig. 2.2 Result of optimization – objective function and the itinerary of the optimization

were different, both optimizations were terminated practically at the same result, after the same number of iterations. Since it is easy to change this parameter it may be verified that the algorithm will have the same performance (in terms of number of iterations) also for different values of initial step lengths within the range [0,1].

A disadvantage of this approach is that every trial step length requires one computation of the function. In this particular case, for the number of parameters

Table 2.1 Optimization results for two different initial step lengths

	x_1	x_2	$f(\mathbf{x})$	x_1	x_2	$f(\mathbf{x})$
Iteration	$\alpha_{IN} = 0.1$			$\alpha_{IN} = 0.5$		
1	1.1000	1.5000	3.4600	1.1000	1.5000	3.4600
2	0.9817	1.3387	2.7560	0.5086	0.6936	0.7398
3	0.7452	1.0162	1.5879	−0.0828	−0.1127	0.0196
4	0.2720	0.3711	0.2117	0.0650	0.0889	0.0121
5	0.1537	0.2098	0.0677	0.0281	0.0385	0.0023
6	0.0355	0.0485	0.0036	−0.0089	−0.0119	0.0002
7	0.0059	0.0082	0.0001	0.0059	0.0082	0.0001
8	−0.0015	−0.0019	0.0001	−0.0015	−0.0019	0.0001

to identify being equal to 2 and with 3 trial steps each iteration included 6 evaluations of function. Considering that the number of iterations was 8, it resulted in total of 48 function evaluations.

Some algorithms that are using inexact line search approach are introducing different criteria that need to be satisfied for the approximated minimization of Eq. 2.10. For example one criterion that can be used is that the step length should give a sufficient reduction of the objective function. This can be measured by the inequality that is found in the literature as *Armijo criterion* ([4, 5]), namely

$$f(\mathbf{x}_k + \alpha_k \mathbf{p}_k) \leq f(\mathbf{x}_k) + c_1 \alpha_k \nabla f_k^T \mathbf{p}_k \tag{2.12}$$

where c_1 is some small scalar constant. The right-hand side of the inequality (2.12) is a linear function of step length and it has a negative slope. It lies above the objective function (or at least it is the case in close proximity of the current iterate) which is guaranteed with c_1 being a small positive number.

It is of course evident that for any sufficiently small value of step length, the Armijo criterion is satisfied. Therefore the goal should be to find the largest possible α_k for which it is satisfied.

A possible implementation of this approach is given in the following MATLAB code. Here the algorithm starts from some initial value of step length and then continues doubling it, until it finds the largest values of α_k that satisfies the inequality. This value is accepted as a reference one for the next iteration. In the case when it violates the inequality (2.12) the step length is divided by 2.

```
**********************************************************
% Optimization algorithm based on line search method.
% Step length identification by Armijo condition

% Setting the options
length=5; % Starting test length for the first iteration
c1=1e-4; % Coefficient for Armijo criterion
pert=0.001; % Perturbation for the first derivatives
guess=rand(2,1)*2-ones(2,1); % Initial vector of parameters
resMIN=1e-6; % The value of residual at the termination

% Optimization cycle
res=10;
iter=0;
while res> resMIN
   itiner(iter+1,1:2)=guess';
   e=exfun(guess);
   itiner(iter+1,3)=e;
   iter=iter+1;
   for i=1:size(guess,1)
       guessp=guess;
       guessp(i)=guessp(i)+pert;
       e1=exfun(guessp);
       grad(i,1)=(e1-e)/pert;
   end
   stpdsc=-grad/norm(grad);
   % Armijo criterion
   foundlenght=0;
   shrt=0; % Total number of eshortening
   while foundlenght==0
       guessTR=guess+length*stpdsc;
       eTR=exfun(guessTR);
       if eTR>e+c1*length*grad'*length*stpdsc;
           length=length/2; % Getting back to previous length
           guessTR=guess+length*stpdsc;
           eTR=exfun(guessTR);
           if eTR<e+c1*length*grad'*length*stpdsc;
               foundlenght=1;
           else
               length=length/2;
               shrt=shrt+1;
           end
       else
           if shrt>0
               break
           end
           length=length*2;
       end
   end
   % Updating values
   guess=guess+length*stpdsc;
   best=eTR;
   res=guess'*guess;
   itiner(iter+1,1:2)=guess';
   itiner(iter+1,3)=eTR;
end
**********************************************************
```

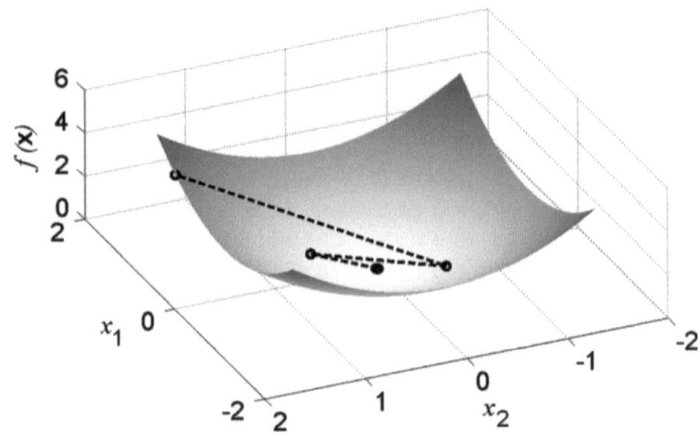

Fig. 2.3 Result of the optimization with Armijo criterion

The remaining two routines are the same as in previous case and can be used also for this optimization. Figure 2.3 visualizes the results of the optimization starting from the same initial point as in previous case, using Armijo criterion for quit large initial step length (equal to 5).

In this case the optimization was terminated after six iterations instead of eight in the previous one visualized in Fig. 2.2. With the adopted strategy however the total number of step lengths that will be tried is not fixed. It means that also the number of function evaluations within the iteration is not limited. Unlike the previous case where in each iteration there were only three candidates, in this one the trials will be repeated until the violation of Armijo criterion. This fact makes the presented implementation more dependent on the initial step length. Repeating the optimization it can be verified that for the initial step length equal to 1, the optimization is terminated after only five iterations, while the starting value of 0.2 involves nine iterations. Some improvements of the approach are of course possible, like for example the one that will put a limitation of the number of increments of the step length in order to avoid waste of computing time in the case when the initial step length is chosen to be quit small.

Satisfying Armijo criterion is not enough by itself to ensure that the algorithm makes reasonable progress since, as previously mentioned, the inequality (2.12) is satisfied for any sufficiently small α_k. This criterion is therefore sometimes combined with the *curvature condition* which requires that the step length satisfies the following condition

$$\nabla f(\mathbf{x}_k + \alpha_k \mathbf{p}_k)^T \cdot \mathbf{p}_k \geq c_2 \nabla f_k^T \cdot \mathbf{p}_k \tag{2.13}$$

where c_2 is some scalar constant with the value within the range $[c_1,1]$. The curvature criterion practically means that the slope of Eq. 2.10 (a function that we are approximately minimizing) at the α_k is c_2 times larger than the initial slope at $\mathbf{x}_k$. Since the initial slope is negative it means that the requirement states that the slope at acceptable step length should be less negative. This is a reasonable requirement since strong negative slope would indicate a possible significant reduction of the objective function by moving further along the chosen direction.

The sufficient decrease and curvature condition are known collectively as *Wolfe conditions*. It represents a strategy for selection of stopping criterion which should provide better trials than Armijo condition alone, but since they require additional gradient computation, as the left-hand side of (2.13) is the derivative of (2.10) at α_k, it may involve even larger number of function evaluations in some cases.

The implementation of the two strategies given here should serve to put in evidence that the step length for the steepest descent plays an important role in overall performance of the optimization algorithm. Both of the two discussed strategies showed some oscillatory behavior around the solution which leads to overall increase in the iterations. Figure 2.3 puts in evidence that already third iteration is located quit close to the result. This suggests that the steepest descend algorithm can be combined with some other method once that it approaches the solution.

2.2.2 Line Search with Newton Direction

Another important search direction is the *Newton direction*. This direction is derived from the second-order Taylor series approximation of the objective function. Using this series to approximate the real objective function around current iterate, and truncating it after the second derivative term will give the following model function

$$f(\mathbf{x}_k + \mathbf{p}_k) \approx m_k(\mathbf{p}_k) = f(\mathbf{x}_k) + \mathbf{p}_k^T \nabla f(\mathbf{x}_k) + \frac{1}{2}\mathbf{p}_k^T \nabla^2 f(\mathbf{x}_k) \cdot \mathbf{p}_k \tag{2.14}$$

Assuming that $\nabla^2 f(\mathbf{x}_k)$ is positive definite, Newton direction can be obtained by finding $\mathbf{p}_k$ that minimizes the model function $m_k(\mathbf{p}_k)$. Finding the first derivatives of m_k with respect to direction and simply setting it to zero will give us the minimizer of the model function, namely

$$\mathbf{p}_k = -\left(\nabla^2 f_k\right)^{-1} \nabla f_k \tag{2.15}$$

Unlike the steepest descent direction, where as we already saw an additional computational effort needs to be made to further identify the step length, Newton direction has a “natural” step length equal to 1. This can be seen from the way the direction is derived. Since it is determined by setting the first derivative of the model function to zero, it means that the Newton direction computed by (2.15) will be *exact* minimizer of model function. It further implies that the reliability of this direction depends on how much the true objective function differs from the model function. If they are almost the same, Newton direction can provide minimizer in one step only.

Let us now consider the same optimization problem as before where objective function is given by (2.11), and let us use the following MATLAB code to solve this problem using Newton direction.

```
*************************************************************
% Optimization algorithm based on line search method.
% That uses Newton direction with the step size equal to 1

% Setting the options
guess=rand(2,1)*2-ones(2,1); % Initial vector of parameters
pert=0.001; % Perturbation for the first derivatives
resMAX=1e-6; % Residual at termination

% Optimization cycle
res=10;
iter=0;

while res>resMAX
   e=exfun(guess);
   itiner(iter+1,1:2)=guess';
   itiner(iter+1,3)=e;
   iter=iter+1;
   % Computing first derivatives
   for i=1:size(guess,1)
       guessp=guess;
       guessp(i)=guessp(i)+pert;
       e1=exfun(guessp);
       grad(i,1)=(e1-e)/pert;
   end
   % Computing Hessian matrix
   HESS=comhess(@exfun,point,pert)
   newton=-inv(HESS)*grad; % Newton direction
   length=1; % Natural length
   guess1=guess+length*newton;
   e1tr=exfun(guess1);
   itiner(iter+1,1:2)=guess1';
   itiner(iter+1,3)=e1tr;
   guess=guess1;
   res=guess'*guess;
end
*************************************************************

*************************************************************
function HESS=comhess(FUNNAME,point,pert)
% Computing the Hessian matrix
pointp=point;
e0=FUNNAME(pointp);
for i=1:size(point,1)
    pointp=point;
    pointp(i)=pointp(i)+pert;
    e1=FUNNAME(pointp);
    pointp(i)=pointp(i)+pert;
```

```
        e2=FUNNAME(pointp);
        Usnd(i)=(e0-2*e1+e2)/(pert^2);
end
% mixed derivative
pointp=point;
pointp(1)=pointp(1)+pert;
U112=FUNNAME(pointp); % term i+1,j
pointp=point;
pointp(2)=pointp(2)+pert;
U121=FUNNAME(pointp); % term i,j+1
pointp=point;
pointp(1)=pointp(1)+pert;
pointp(2)=pointp(2)+pert;
U1121=FUNNAME(pointp); % term i+1,j+1
mixed=1/pert*((U1121-U112)/pert-(U121-e0)/pert);
HESS(1,1)=Usnd(1);
HESS(1,2)=mixed;
HESS(2,1)=mixed;
HESS(2,2)=Usnd(2);
************************************************************
```

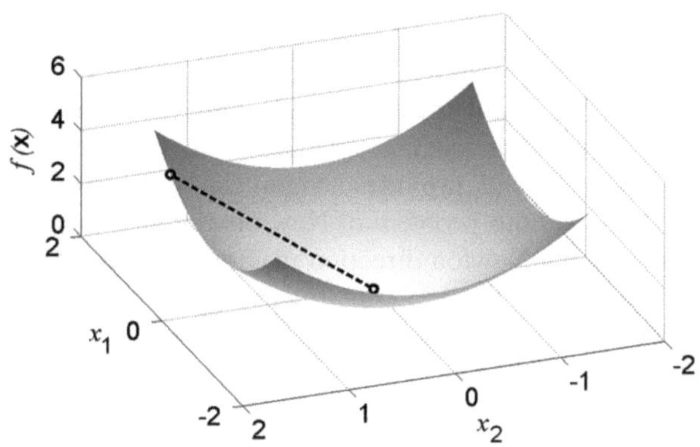

Fig. 2.4 Result of the optimization with Newton direction – convergence after one step only

Note that the second routine given in the listing is a MATLAB function used to compute Hessian matrix. This function will be also used later in other optimization algorithms.

The optimization problem was solved starting from the same initial point as in previous cases and the result is visualized in Fig. 2.4.

The real power of Newton direction is evidenced in this example since the algorithm converged in one iteration only. Unlike steepest descend approach, which selects the direction, and then, depending on the strategy of the step length search, the algorithm may take couple of iterations to reach the solution, Newton direction immediately finds the step that exactly minimizes the model function. In this case the objective function was practically the same as the model function, and so the optimization terminated after only one iteration.

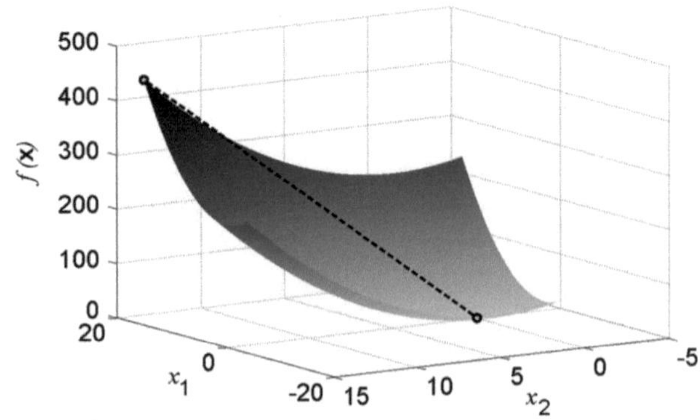

Fig. 2.5 Newton direction optimization starting far from the solution

The example considered here have simple, smooth and convex objective function that is suitable to be modeled by second-order Taylor series. It can be shown that it is the case, even farther from the solution. Figure 2.5 shows the result of the optimization when the initialization point was quit far from the solution being equal to $[15,15]^{\mathrm{T}}$. Still however, the Newton direction provided the solution in one iteration.

In this code, derivatives are computed by finite differences. This represents a disadvantage in terms of computing times. However, if the objective function, like in this case can, with acceptable accuracy, be represented by a second-order model, this drawback is acceptable since, as we saw, the Newton direction, with respect to steepest descend, provides a significant reduction in the number of iterations.

More serious problem of Newton direction is that, when Hessian is not positive definite, the Newton direction may not be defined, or if it is defined it may not be a descending direction. In order to overcome this problem there are different approaches in which the Hessian matrix is modified in order to make it positive definite and thereby yield a descent direction. These problems will arise in the situations where the objective function is more complicated than the one studied in this example. Therefore in the following pages the behavior of both steepest descent and Newton direction method will be analyzed on the least squares type objective function.

2.2.3 *Line Search in Least Squares Problems*

Let us consider the following analytical function

$$y(t) = t \cdot \sin(t) - \sqrt{t} \tag{2.16}$$

which, at the domain $t \in [0,3]$ is visualized in Fig. 2.6. Now let us imagine that this expression represents the distribution of a certain physical value (here y) at some

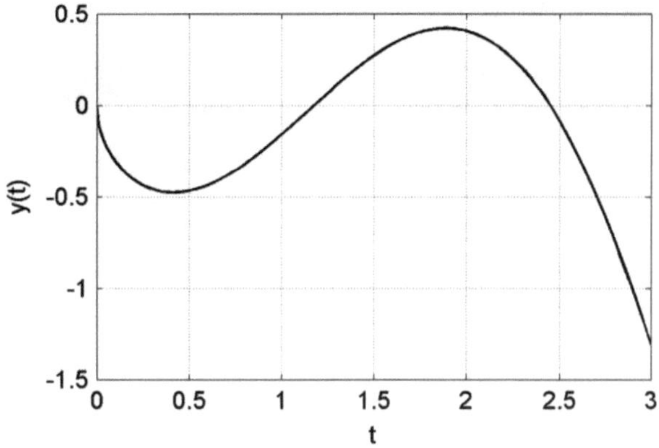

Fig. 2.6 Graph of analytical function (2.16), used as "target" function

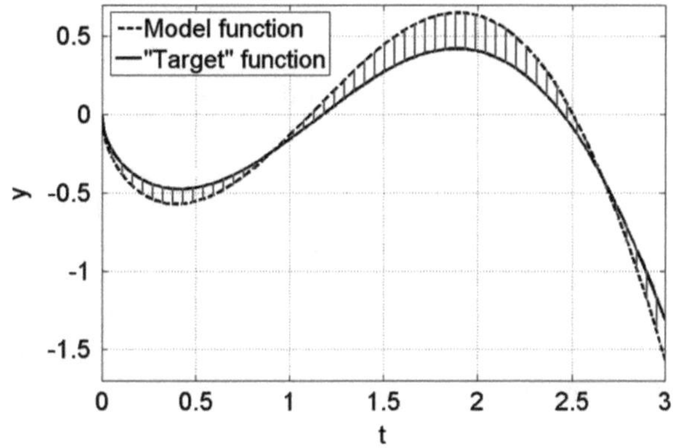

Fig. 2.7 Model and "target" function with the distances between the two curves over some grid of points

time or spatial coordinate t. Let us further suppose that we have some model function that represents a computed counter-part of this experiment given by

$$y^{COM}(t, \mathbf{x}) = x_1 \cdot t \cdot \sin(t) - \sqrt{x_2 \cdot t} \tag{2.17}$$

The model function is function of two parameters collected in vector $\mathbf{x}$.

Let us assume that we would like to identify the two parameters x_1 and x_2 for which the model function (2.17) will match the "target" one given by (2.16). Obviously, the target function exists within the family of model functions and it is obtained for $\mathbf{x} = [1,1]^T$. In order to design the inverse analysis procedure that will identify these two parameters the first step is to build the objective function that will quantify the discrepancy between the target and model function. This can be done using the discrepancy function in the least squares form that will represent a summation of squares of differences between the two function values for some grid over t, as shown in Fig. 2.7. The objective function will have the following form

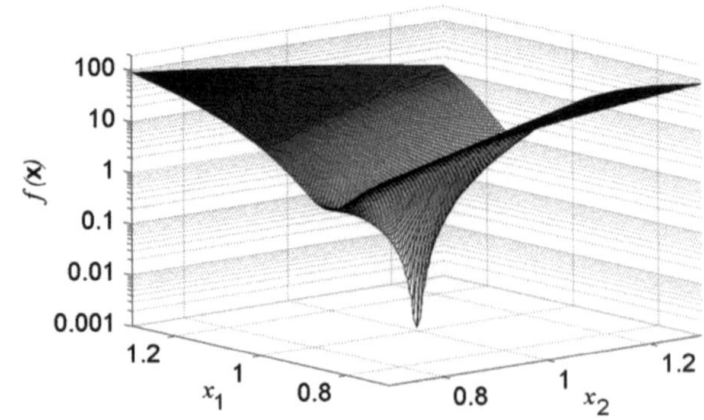

Fig. 2.8 Discrepancy function as summation of squares of differences between the two curves

$$f(\mathbf{x}) = \frac{1}{2}\sum_{i=1}^{N_p}[y(t_i) - y(t_i, \mathbf{x})]^2 \tag{2.18}$$

where N_p is the number of grid points over t used to compute the distance between the two curves. It should be mentioned that this is a typical formulation in any situation where parameterized models are used to predict some phenomenon. In such cases the discrepancy between model prediction and observed behavior is usually measured by the function of the type given by Eq. 2.18. In this case we simplified the situation by considering as observation "clean" analytical data so that the exact solution will exist within our model, in order to study only the behavior of different optimization approaches in the situation in which we know exactly what the solution is.

The objective function defined in this way has more complicated shape than in previously studied case and its form is visualized in Fig. 2.8.

Following MATLAB code can be used to minimize objective function (2.18) using steepest descend approach. The code is practically the same as previously given for optimization of function (2.11) except that here additional lines for different convergence criteria are inserted. Here the optimization is terminated not only in the case when residual is smaller than some prescribed value, but also if the changes of the parameters are smaller than certain value and if the number of iterations is larger than given number `NUMIT`. This is a common practice in more complicated optimizations since there, a single criterion is not enough and may even lead to the optimization that will never be terminated.

```
*************************************************************
% Optimization algorithm based on Line search with steepest
% descend, with Armijo condition
clear

% Setting the options
minchg=1e-5; % Min change in parameters between two iterations
guess=rand(2,1)*2;
minRES=1e-6; % Residual at termination
NUMIT=70; % Max number of iteration allowed
pert=0.001; % Perturbation for the derivatives
length=0.01; % Starting test length for the first iteration
c1=1e-2; % Coefficient for Armijo criterion

% Optimization cycle
res=10;
iter=0;
while res>minRES
   itiner(iter+1,1:2)=guess';
   e=funLSQ(guess);
   itiner(iter+1,3)=e;
   iter=iter+1;
   for i=1:size(guess,1)
       guessp=guess;
       guessp(i)=guessp(i)+pert;
       e1=funLSQ(guessp);
       grad(i,1)=(e1-e)/pert;
   end
   stpdsc=-grad/norm(grad);
   % Armijo criterion
   foundlenght=0;
   while foundlenght==0
       guessTR=guess+length*stpdsc;
       eTR=funLSQ(guessTR);
       if eTR>e+c1*length*grad'*length*stpdsc;
           length=length/2; % Getting back to previous lenght
           guessTR=guess+length*stpdsc;
           eTR=funLSQ(guessTR);
           if eTR<e+c1*length*grad'*length*stpdsc;
               foundlenght=1;
           else
               length=length/2;
           end
       else
           length=length*2;
       end
   end
   % Updating values
   guess=guess+length*stpdsc;
```

```
    best=eTR;
    res=best;
    itiner(iter+1,1:2)=guess';
    itiner(iter+1,3)=best;
    % Checking convergence options
    if abs(itiner(iter+1,1)-itiner(iter,1))<minchg ||
         abs(itiner(iter+1,2)-itiner(iter,2))<minchg;
      res=0;
    end
    if iter> NUMIT % Terminated after the iteration reach NUMIT
        res=0;
    end
end
***********************************************************

***********************************************************
function e=funLSQ(x);
exper=load('lsqexp.txt');
A=x(1);
B=x(2);
cnt=0;
for i=0:0.01:3
    cnt=cnt+1;
    y(cnt)=A*sin(i)*i-sqrt(B*i);
end
e1=y'-exper(:,2);
e=e1'*e1;
***********************************************************
```

The second MATLAB code is used to compute the discrepancy function between model curve and "experimental" one. Here the "experimental" curve is placed in the file `lsqexp.txt` that is actually the graph of Eq. 2.16 (Fig. 2.6).

The result of the optimization is visualized in Fig. 2.9. Setting the total allowed number of iterations to 70 and other convergence criteria quite strict (minimum change in parameters 1E-5 and minimum residual at the termination 1E-6) the optimization terminates after reaching 70th iteration, at parameter values of $\mathbf{x} = [0.998, 0.995]^T$.

This objective function is a typical example which demonstrates that the steepest descend is not the best direction to go along. As it can be observed from the figure,

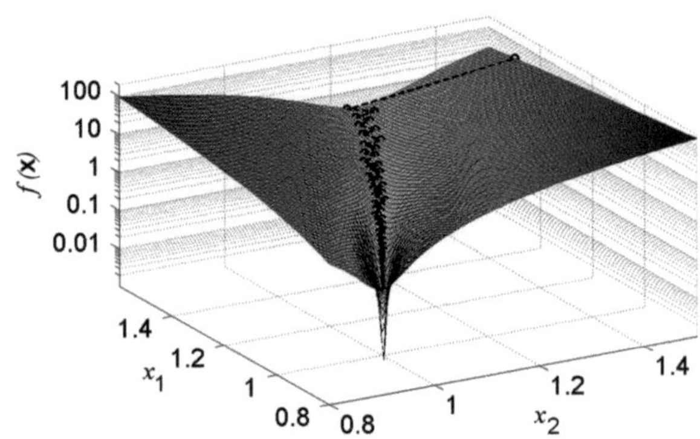

Fig. 2.9 Itinerary of steepest descend optimization of least squares discrepancy function given by Eq. 2.18

practically in all intermediate points within the optimization the steepest descend direction is not matching the direction of global minimum. Therefore, only the small steps can be achieved with which the optimization slowly approaches the minimum, which causes large number of iterations. Since the reason for poor performance is the direction itself, no improvements can be obtained with different strategies for step length selection.

The same problem can be solved using Newton direction. For this purpose a previously given MATLAB code with the same modifications as those for the steepest descend can be used.

The Newton direction approach turns out to be much more effective. Using the same tolerances as for the steepest descend the algorithm converges after only seven iterations but practically already the forth one is in the vicinity of the solution. Figure 2.10 visualizes the itinerary and Table 2.2 lists all the parameter values and corresponding objective function values.

The last example proved the superiority of the Newton direction in the cases when it can be computed, and when it is a descending direction. However, using this direction in the line search approach is not that effective all the time. Apart from being computationally more expensive with respect to steepest descend by involving the computation of second derivatives, there are two main problems connected with the successful implementation of Newton direction.

The first one is that, as already evidenced, the Newton direction minimizes exactly the model function which is a quadratic form of a real objective function around the current iterate. This is a good approximation in the vicinity of the current iterate but by going farther it starts to worsen. It may happen, when the minimizer of the model function is far from the current iterate, that it will be a poor minimizer of the real

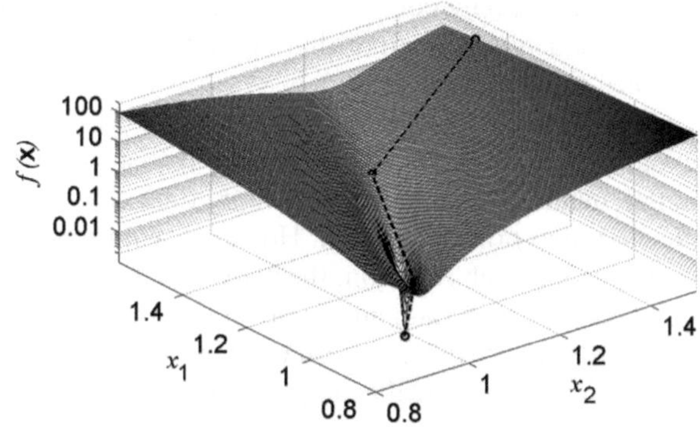

Fig. 2.10 Itinerary of Newton direction optimization of least squares discrepancy function given by Eq. 2.18

Table 2.2 Optimization results for Newton direction and LS function

Iteration	x_1	x_2	$f(\mathbf{x})$
1	1.5000	1.5000	38.3040
2	1.1337	1.2926	0.9928
3	0.9847	0.9450	0.0975
4	0.9966	0.9923	0.0008
5	0.9994	0.9987	0.0000
6	0.9994	0.9988	0.0000
7	0.9994	0.9988	0.0000

function, or it can even increase its value. This can be encountered especially in the cases when iterate is quite far from the global minimum point of the real objective function. An alternative approach that overcomes this problem is *Trust Region* approach and it will be discussed in more details further in this chapter.

The second problem is connected with the fact that Hessian matrix may not be positive definite which means that the Newton direction computed by Eq. 2.15 may not be a descent direction. This problem may be solved in the line search context by modifying the Hessian matrix in order to force it to be positive definite. Unlike the first abovementioned problem, where the difference between the model function and the real one can be verified only after the value of the objective function is evaluated for the next iteration (computed by adding Newton direction to the current iterate) tackling of the second problem is not that time consuming. After the Hessian matrix is computed, if it runes out not to be positive definite, there is no need to use this step to evaluate the objective function as it is not guaranteed that it will reduce the objective function. Instead, an appropriate modification of Hessian matrix can be adopted in order to yield a descending direction. Implementing such modification in the algorithm is not computationally "expensive", as it is not involving any further evaluation of the objective function.

The Hessian matrix modification is usually obtained by adding either a positive diagonal matrix or full matrix to the true Hessian. However the step computed by modified Hessian matrix looses its "natural" length of 1, as it is not anymore the exact minimizer of the model function. If the modified Hessian is positive definite, the direction computed using it will be a descending one but the step length may have to be adjusted to ensure a reasonable decrease of the objective function like in the line search approach with steepest descent direction. Usually in the line search algorithms that use modified Newton direction, the step length 1 is tried first and then some stopping conditions like those discussed previously are checked. If they are not satisfied, the algorithm should modify the step length, otherwise it may proceed to the next iteration.

Even though the fact that Hessian is not positive definite can be verified a priori avoiding a useless objective function evaluation, still the need to evaluate the step length makes the approach a bit more time consuming with respect to the classical Newton direction line search. However, with this modification, as it is usually not performed at each iteration, the Newton direction line search usually tends to keep good convergence rate, and in general is more effective than the steepest descend.

In order to illustrate the effect of modification of Hessian matrix let us consider the following numerical example. Let us assume that the target distribution of some physical value on the domain $t \in [0, 3]$, that in this context represents "experimental" data, is given by the following equation (Fig. 2.11)

$$y = 2.5 \cdot \sin t + e^{\cos(2.5t)} \tag{2.19}$$

Like in previously discussed case, let us suppose that the computed counter-part of the "experimental" values is given with the following model function

$$y^{COM}(t, \mathbf{x}) = x_1 \cdot \sin t + e^{\cos(x_2 t)} \tag{2.20}$$

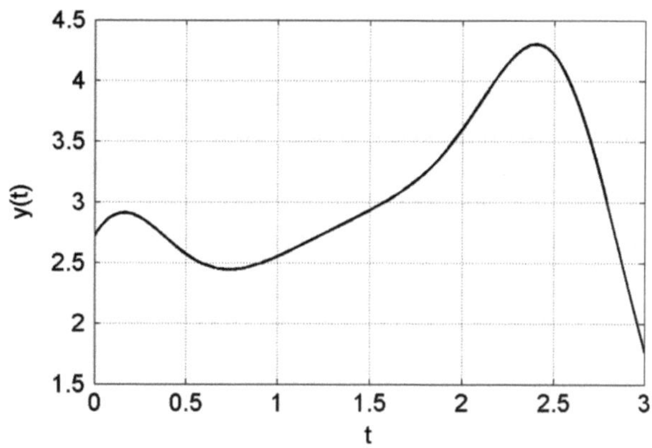

Fig. 2.11 Graph of analytical function (2.19), used as "target" function

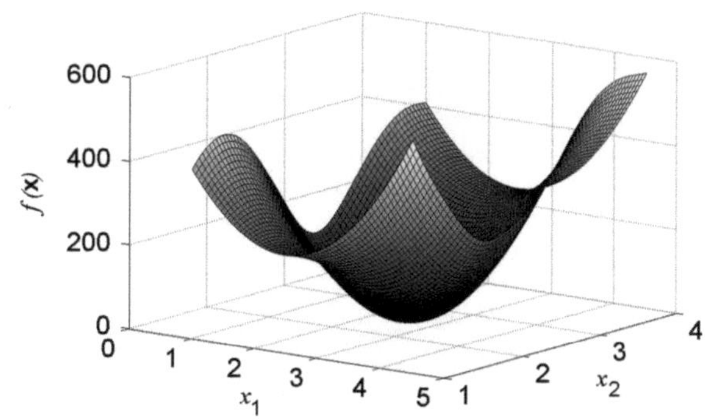

Fig. 2.12 The objective function to be minimized

which depends on two parameters collected in vector **x**.

The goal is to identify the two model parameters by constructing a procedure that will minimize the discrepancy function in the least squares form, like the one given by Eq. 2.18. The objective function to be minimized is visualized in Fig. 2.12.

The following MATLAB code can be used to minimize this objective function using both classical Newton direction and modified one. Here, the modification is performed by adding a diagonal matrix that for this particular case turned out to be good enough. In general, a wide variety of types of modifications can be used, and the reader is referred to the reference [2] for more details on this topic. As mentioned earlier, after the Hessian modification, it is also needed to check for the step length, as it cannot be anymore a priori assumed as equal to 1. In the code given here the first tried value was 0.5, and then it is further reduced if necessary. The objective function studied here is characterized with large number of local minima when the parameters are changing in wider ranges. In order to avoid trapping into one of those, as the tackling of local minima is not an issue at this stage, the goal was to avoid large steps along the descending directions. Therefore the maximum allowed step length for modified Newton direction was assumed to be 0.5 unlike the common practice where it is usually 1.

In order to evidence the advantage of least squares method in terms of easy approximation of Hessian matrix by computing only first derivative, the following code has an option either to compute full Hessian matrix, or to approximate it by using just Jacobian (Eq. 2.7). This option is set by the variable `Hessapp`. By attributing value 1 to it, Hessian approximation will be used, otherwise, also second derivatives will be computed. Both of the functions used for this computation are given at the bottom of the code. Note that with respect to the case presented in Sect. 2.2.2, also the routine for computing Hessian matrix is modified in order to take into account function that gives vector-value (i.e. vector of residuals).

```
***********************************************************
% Line search algorithm with Newton direction with possibility
% of Hessian matrix modification and approximation
clear

% Setting the options
minchg=1e-4; % Minimum change in parameters
MAXIT=30; % Maximum allowed number of iterations
guess=[1.5;2.9];
pert=1e-6; % Perturbation for the first derivatives
res=10;
HessMod=0; % Indication for Hessian modification
Hessapp=0; % Approximating Hessian matrix

% Optimization cycle
iter=0;
while res>1e-6
  eV=funLSQ1(guess);
  e=0.5*eV'*eV;
  itiner(iter+1,1:2)=guess';
  itiner(iter+1,3)=e;
  iter=iter+1;
  for i=1:size(guess,1)
      guessp=guess;
      guessp(i)=guessp(i)+pert;
      eV=funLSQ1(guessp);
      e1=0.5*eV'*eV;
      grad(i,1)=(e1-e)/pert;
  end
  guessp=guess;
  eV=funLSQ1(guessp);
  e0=eV'*eV;
  if Hessapp==1
      HESS=comhessapp(@funLSQ1,guess,pert);
  else
      HESS=comhess(@funLSQ1,guess,pert);
  end
  newton=-inv(HESS)*grad;
  length=1;
  lambdas=eigs(HESS);
  if HessMod==0
      lambdas=abs(lambdas); %Trick to avoid Hessian
modification
  end
% Checking if Hessian is positive definite
  if lambdas(1)>0 && lambdas(2)>0
      hessmod=0;
  else
      coeff=mean(abs(lambdas));
```

```
        posdef=0;
        while posdef<1
            HESSm=HESS+coeff*eye(2); % Adding diagonal matrix
            hessmod=1; % Indication of modified HESSIAN
            lmb=eigs(HESSm);
            if lmb(1)>0 && lmb(2)>0
                posdef=1;
            else
                coeff=coeff*1.5;
            end
        end
    end
% Computing next iterate
    if hessmod==0
        guess1=guess+length*newton;
        eV=funLSQ1(guess1);
        e1tr=0.5*eV'*eV;
    else
        foundstep=0;
        length=0.5;
        while foundstep==0
            mnewton=-inv(HESSm)*grad; % Modified Newton
direction
            guess1=guess+length*mnewton;
            PQN=length*mnewton; % Step for quasi newton
direction
            modf=e0+PQN'*grad+PQN'*HESS*PQN;
            if modf<e0
                foundstep=1;
            else
                length=length/1.5;
            end
        end
        eV=funLSQ1(guess1);
        e1tr=0.5*eV'*eV;
    end
    itiner(iter+1,1:2)=guess1';
    itiner(iter+1,3)=e1tr;
    guess=guess1;
    res=guess'*guess;
    if iter>MAXIT % Terminated after the iteration reaches MAXIT
        res=0;
    end
% Checking convergence options
    if abs(itiner(iter+1,1)-itiner(iter,1))<minchg ||
abs(itiner(iter+1,2)-itiner(iter,2))<minchg;
        res=0;
    end
end
*************************************************************
```

```
************************************************************
function e=funLSQ1(x);
exper=load('lsqexp1.txt');
A=x(1);
B=x(2);
cnt=0;
for j=0:0.01:3
    cnt=cnt+1;
    y(cnt)=A*sin(j)+exp(cos(B*j));% *j+B*j^2;
end
e1=y'-exper(:,2);
************************************************************

************************************************************
function HESS=comhess(FUNNAME,point,pert)
% Computing the Hessian matrix
pointp=point;
eV=FUNNAME(pointp);
e0=0.5*eV'*eV;
for i=1:size(point,1)
    pointp=point;
    pointp(i)=pointp(i)+pert;
    eV=FUNNAME(pointp);
    e1=0.5*eV'*eV;
    pointp(i)=pointp(i)+pert;
    eV=FUNNAME(pointp);
    e2=0.5*eV'*eV;
    Usnd(i)=(e0-2*e1+e2)/(pert^2);
end
% mixed derivative
pointp=point;
pointp(1)=pointp(1)+pert;
eV=FUNNAME(pointp); % term i+1,j
U112=0.5*eV'*eV;
pointp=point;
pointp(2)=pointp(2)+pert;
eV=FUNNAME(pointp); % term i,j+1
U121=0.5*eV'*eV;
pointp=point;
pointp(1)=pointp(1)+pert;
pointp(2)=pointp(2)+pert;
eV=FUNNAME(pointp); % term i+1,j+1
U1121=0.5*eV'*eV;
mixed=1/pert*((U1121-U112)/pert-(U121-e0)/pert);
HESS(1,1)=Usnd(1);
HESS(1,2)=mixed;
HESS(2,1)=mixed;
HESS(2,2)=Usnd(2);
************************************************************
```

```
************************************************************
function HESS=comhessapp(FUNNAME,point,pert)
% Computing the Hessian matrix
pointp=point;
e0=FUNNAME(pointp);
% Computing Jacobian
for i=1:size(point,1)
    pointp=point;
    pointp(i)=pointp(i)+pert;
    e1=FUNNAME(pointp);
    J(:,i)=(e1-e0)/pert;
end
% Hessian approximation
HESS=J'*J;
************************************************************
```

In the code listed above, whether, in each iteration, a true Newton direction is used, or it is modified in order to enforce descending step, is chosen by variable `HessMod`.

The optimization problem was solved two times using as initialization point $\mathbf{x} = [1.5, 2.9]^{\mathrm{T}}$. First time it was solved without Hessian modification. The result is visualized in Fig. 2.13. Upper graph shows how the parameters are changing during the iterations, while the down one shows the changes in the objective function. It may be observed that from second to third iteration there is a large jump in the value of the objective function. It is due to the fact that the Hessian matrix computed at this iteration is not positive-definite. At this point, the eigenvalues of Hessian matrix are equal to: $\lambda_1 = 251.7$ and $\lambda_2 = -8.7$. As a consequence to that, the computed Newton step is not descending even for the model function. Without modifying Hessian matrix, and by accepting this step, third iteration goes far from the solution, and eventually finishes in the local minimum.

The same problem can be solved using given code by setting the variable `HessMod` equal to 1, which will introduce the modification of Hessian matrix in the cases when it is not positive definite. The results are visualized in Fig. 2.14. It may be observed that in this case there is a continuous reduction of the function. In the second iteration the Hessian matrix was modified and the step was descending. After this iteration there was no need to perform any modification and the algorithm finished with true Newton directions showing a very fast rate of convergence and finding the parameters after only five iterations.

Finally, in order to verify that for least squares problems, especially not very far from the solution, also Hessian approximation can be used almost with the same efficiency the problem is once again solved but having set variable `Hessapp` equal to 1. Figure 2.15 visualizes the results, from which it may be confirmed that in this particular case there was no degradation in performance of the optimization

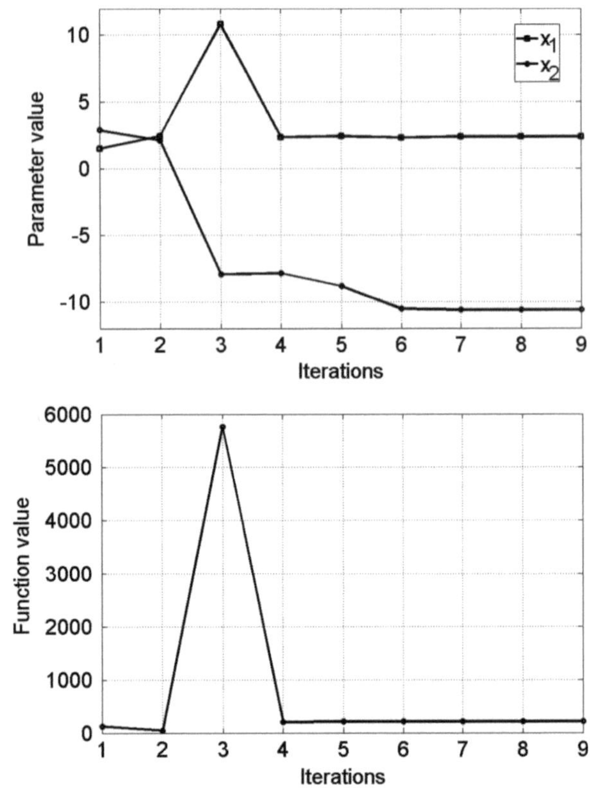

Fig. 2.13 Results of optimization without modifying Hessian matrix

algorithm even when Hessian approximation is used. This particular feature can give a significant savings in computing times when the function evaluation involves possibly time consuming FE analyses.

The same problem can be solved using previously given MATLAB code for steepest descend. If for example the one with three trial steps is employed, for this particular case it turns out to be more effective than in the optimization of objective function given by Eqs. 2.16–2.18, and it converges after 24 iterations. The results are given in Fig. 2.16. It can be noticed that the algorithm with modified Newton direction is overall more effective than steepest descend, as it involves less evaluations of function.

Previous example showed that, even though the use of true Newton direction can lead sometimes to steps with increase of the objective function, this malfunction can be overcame relatively easy by implementing Hessian modification. With this approach the algorithm preserves a fast convergence rate and in most of the cases can be more effective than the steepest descend direction.

Another problem connected with Newton direction is that in some cases, especially far from the solution the model function can be quit different from the real one. It may result that, the Newton direction will provide a step that successfully minimizes the model function but which is not minimizing enough or even not at all

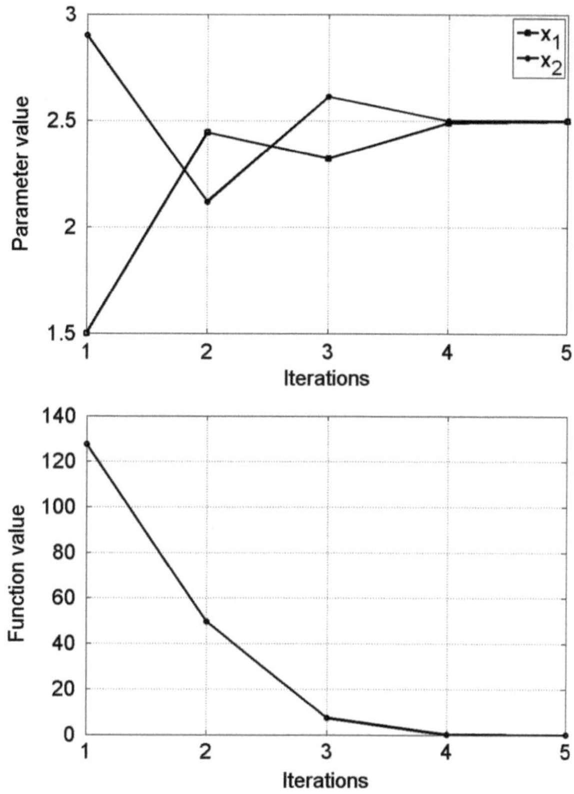

Fig. 2.14 Results of optimization with Hessian modification

the real function. To tackle this problem a logical approach is to use the model function only in the vicinity of the current iterate. This is approach adopted in the *Trust Region* method.

2.3 Trust Region

Both line search with Newton direction and trust region are using the quadratic model of objective function, but they differ however in the way they make the use of this model. Line search starts by fixing the direction and then identifies an appropriate distance, namely the step length. Trust region on the other hand, first chooses the maximum distance – the trust region radius Δ_k – and then seeks both the direction and the step length that makes the best possible improvement of the function inside the trust region. This approach helps dealing with the situations when the quadratic model is quite different from the actual function as it may occur far from the solution. The model function will be close to the objective function in the vicinity of the current point, so restricting the minimization just to that area is a reasonable strategy.

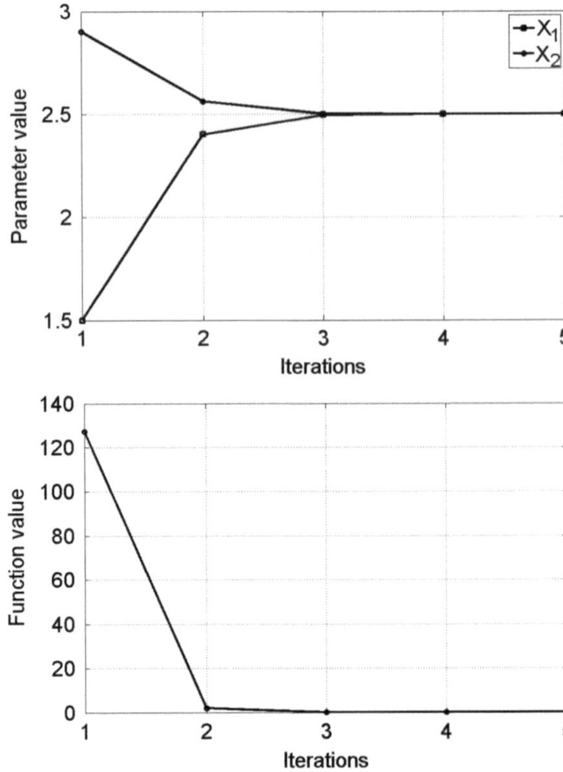

Fig. 2.15 Results of optimization with Hessian approximation

One of the main issues of the trust region approach, that to a large extent determines the success and the performance of this algorithm, is the decision strategy of how large the trusted region should be. Allowing it to be too large can make the algorithm facing the same problem as the classical Newton direction line search, when the minimizer of model function is quite far from the minimizer of the actual objective function. On the other hand using too small region the algorithm misses an opportunity to take a substantial step that could move it much closer to the solution. Some approaches on how to control this important issue will be discussed later.

Each step in the trust region algorithm is obtained by solving the sub-problem defined by

$$\min m_k(\mathbf{p}_k) = f(\mathbf{x}_k) + \mathbf{p}_k^T \nabla f(\mathbf{x}_k) + \frac{1}{2}\mathbf{p}_k^T \nabla^2 f(\mathbf{x}_k) \cdot \mathbf{p}_k$$
$$\|\mathbf{p}_k\| \leq \Delta_k \tag{2.21}$$

where Δ_k is the trust region radius.

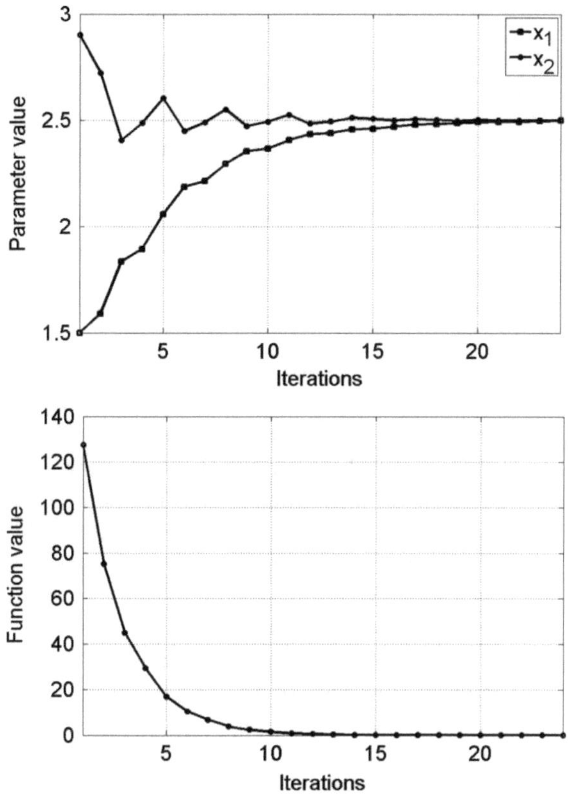

Fig. 2.16 Results of optimization with steepest descend and three trial steps

From previous discussion it may be observed that there are essentially two important parts of any trust region algorithm. The first one, already anticipated, is the way the algorithm controls the radius of trust region. The second one is how efficiently it solves the sub-problem defined by (2.21). In what follows it will be shown how both of these problems are tackled and implemented within different optimization algorithms.

2.3.1 Trust Region Algorithm Based on Cauchy Point

As previously demonstrated on steepest descend line search approach, the algorithm can have global convergence characteristic even when the optimal step length is not used at each iteration. A similar reasoning applies also in trust region methods. Although it is convenient to find optimal solution of the sub-problem (2.21), for global convergence it is enough to find an approximate solution that lays within the trust region and gives a sufficient reduction of the model function.

This approach can be obtained by the use of so-called *Cauchy point*, which is a point that minimizes the model function along the steepest descent direction subjected to the trust region bound. However, the presence of model function makes the minimization of sub-problem a lot easier than in the steepest descend line search procedure. After choosing the search direction, the problem becomes one-dimensional but in the Cauchy point trust region algorithm it can be solved more effectively than in the steepest descent line search since here it is applied to the model function. Therefore, the evaluation of the function is computationally inexpensive.

Considering that the step length anyhow cannot exceed the trust region radius, a simple strategy for finding the Cauchy point can be used. First the value of model function can be computed for the values of steps equal to radius Δ_k, $0.75\Delta_k$ and $0.5\Delta_k$. Based on these values it is easy to see in which zone the minimum is located. In the second iteration algorithm computes the value of model function for the intermediate point between the two boundary points and once again identifies the zone of the solution. While the algorithm propagates the zone of minimum shrinks and already after couple of iterations it is reasonably close to the solution. After the last iteration one of the two boundary points with lower value of model function is taken as a solution. Figure 2.17 shows schematically one case of the Cauchy point identification based on this strategy.

As it can be seen from the figure, in fourth iteration the Cauchy point is identified with an error of about $0.015\Delta_k$. However, considering that these iterations involve only the computation of model function it is computationally inexpensive to repeat this procedure for additional couple of iterations that would practically find the exact minimizer of the model function along steepest descent direction.

Apart from the adopted strategy to solve the sub-problem an important issue of any trust region algorithm is the way it controls the trust region radius Δ_k. In most practical algorithms the choice of size of the region is performed according to performance of the algorithm during previous iteration. In particular the choice is

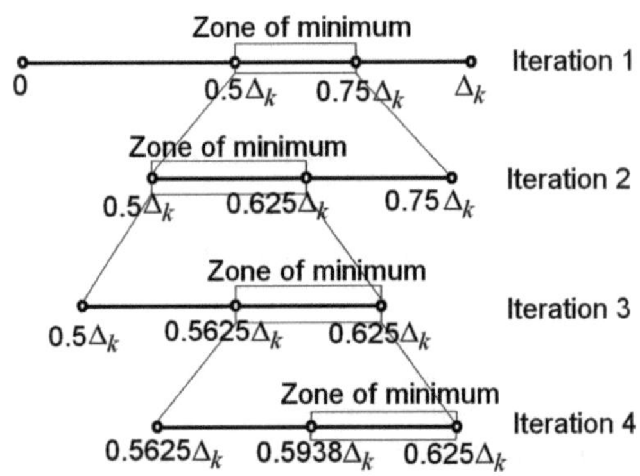

Fig. 2.17 Finding a Cauchy point approximately with four iterations

made based on the agreement between the model function and the objective function. This agreement is quantified by the following ratio

$$\rho_k = \frac{f(\mathbf{x}_k) - f(\mathbf{x}_k + \mathbf{p}_k)}{m(0) - m(\mathbf{p}_k)} \tag{2.22}$$

The denominator represents what is called *predicted reduction,* while the numerator is called *actual reduction* obtained with the computed step on the objective function. Therefore, the closer this ratio is to 1, the better agreement between the objective function and the model is. Since predicted reduction will always be a positive number, as this is a condition in solving the sub-problem, if the ratio (2.22) turns out to be negative, it means that there was a significant disagreement between model function and the objective function. In such case the step should be rejected since it increases the objective function and the trust region should be shrank.

The ratio (2.22) can be computed only after an additional evaluation of the objective function, which therefore represents a wasted computing time if the step is rejected. To avoid this possible inconvenience, this ratio is used as an indicator of how accurately model function describes the real one. Therefore, in most practical algorithms, if the ratio (2.22) is close to zero, the trust region radius should be reduced for the next iteration in order to avoid possible step rejection. On the other hand if it is close to 1 it is a signal that the trust region can be enlarged for the next iteration and therefore allow for possible larger and more ambitious steps. In general however, the trust region radius is not enlarged if the minimizer is found strictly inside the region as in such case it is not interfering with the progress of the algorithm.

Implementation of trust region algorithm that uses Cauchy point approach to solve the sub-problem is given in the following MATLAB code. The listing contains two separate routines – the main one, and an additional function that computes Cauchy point using the strategy schematically presented in Fig. 2.17. The value of initial trust region radius can be selected together with other options. As addition to these also the previously given MATLAB function `funLSQ1` is used. During the optimization if the ratio (2.22) is smaller than 0.2 the algorithm reduces the step by 20%. In the case when this ratio is larger than 0.6 the radius is enlarged by 20%. Otherwise, it remains unaltered.

```
************************************************************
% Trust region algorithm with Cauchy point approach for
% sub-problem
clear

% Setting the options
minchg=1e-4; % Minimum change in parameters
MAXIT=30; % Maximum allowed number of iterations
guess=[1.7;2.9];
pert=1e-6; % Perturbation for the first derivatives
res=10;
TRrad=0.8;

% Optimization cycle
iter=0;
while res>1e-6
  eV=funLSQ1(guess);
  e0=0.5*eV'*eV
  itiner(iter+1,1:2)=guess';
  itiner(iter+1,3)=e0;
  iter=iter+1;
  for i=1:size(guess,1)
      guessp=guess;
      guessp(i)=guessp(i)+pert;
      eV=funLSQ1(guess);
      e1=0.5*eV'*eV;
      grad(i,1)=(e1-e0)/pert;
  end
  % Computing Hessian matrix
  HESS=comhess(@funLSQ1,guess,pert);
  % Computing steepest descent direction
  stpdsc=-grad/norm(grad);
  % Determining TR step
  accepted=0;
  while accepted<1
      % Finding Cauchy point
      pc=cauchypnt(e0,stpdsc,grad,HESS,TRrad);
      predred=-(pc'*grad+0.5*pc'*HESS*pc);
      guess1=guess+pc; % Next iterate
      eV=funLSQ1(guess);
      e1tr=0.5*eV'*eV;
      actualred=e0-e1tr;
      ratio=actualred/predred;
      if ratio<0
          TRrad=TRrad/1.2;
      else
          accepted=1;
          if ratio<0.2
              TRrad=TRrad/1.2;
          end
```

```
        if ratio>0.6
            TRrad=TRrad*1.2;
        end
    end
  end
  itiner(iter+1,1:2)=guess1';
  itiner(iter+1,3)=e1tr;
  guess=guess1;
  res=e1tr;
  % Checking convergence options
  if iter>MAXIT % Terminate if reaching MAXIT
    res=0;
  end
  if abs(itiner(iter+1,1)-itiner(iter,1))<minchg ||
abs(itiner(iter+1,2)-itiner(iter,2))<minchg;
      res=0;
  end
end
**********************************************************

**********************************************************
function pc=cauchypnt(e0,stpdsc,grad,HESS,TRrad)
% Function that computes Cauchy point

lenC1=TRrad; % point 1
lenC2=0.75*TRrad; % point 2
lenC3=0.5*TRrad; % point 3
pc1=lenC1*stpdsc;
pc2=lenC2*stpdsc;
pc3=lenC3*stpdsc;
modfun1=e0+pc1'*grad+0.5*pc1'*HESS*pc1;
modfun2=e0+pc2'*grad+0.5*pc2'*HESS*pc2;
modfun3=e0+pc3'*grad+0.5*pc3'*HESS*pc3;
if modfun1<modfun2
    lenC(1)=lenC1; % result between point 1 and 2
    lenC(2)=lenC2;
else
    if modfun3<modfun2
        lenC(1)=lenC2; % result between point 2 and zero
        lenC(2)=0;
    else
        if modfun1<modfun3
            lenC(1)=lenC1; % result between point 1 and 2
            lenC(2)=lenC2;
        else
            lenC(1)=lenC2; % result between point 2 and 3
            lenC(2)=lenC3;
        end
    end
end
```

```
ic=0;
for NR=1:8
    ic=ic+2;
    pc1=lenC(ic-1)*stpdsc;
    pc2=lenC(ic)*stpdsc;
    modfun1=e0+pc1'*grad+0.5*pc1'*HESS*pc1;
    modfun2=e0+pc2'*grad+0.5*pc2'*HESS*pc2;
    lenC(ic+1)=0.5*(lenC(ic)+lenC(ic-1));
    if modfun1<modfun2
        lenC(ic+2)=lenC(ic-1);
    else
        lenC(ic+2)=lenC(ic);
    end
end
IC=size(lenC,2);
pc1=lenC(IC-1)*stpdsc;
pc2=lenC(IC)*stpdsc;
modfun1=e0+pc1'*grad+0.5*pc1'*HESS*pc1;
modfun2=e0+pc2'*grad+0.5*pc2'*HESS*pc2;
if modfun1<modfun2 % Identifying Cauchy point
    pc=pc1;
else
    pc=pc2;
end
**********************************************************
```

Using given routines optimization problem defined by Eqs. 2.19 and 2.20 can be solved. In order to demonstrate the capability of trust region approach to tackle the problems when quadratic model function is quite different from the objective function we can solve the optimization problem starting from the point $[1.7, 2.9]^T$.

The same optimization problem can be solved by using routines given before that use Newton line search approach with Hessian modification. Starting from the same point we can notice that in the second iteration algorithm arrived to the parameter values of $[2.45, 2.05]^T$. Hessian matrix at this point is positive definite so the algorithm proceeds with the exact Newton step. Computed Newton step for this iteration is $[-0.747, 1.552]^T$. Even though this step minimizes the model function, it produces the increment in the actual objective function due to the significant difference between the model and objective function. It represents a classical example where Newton line search method with Hessian modification is ineffective resulting in increase of the objective function, as this increase results not because the Hessian matrix is not positive definite, but because the minimizer of the model function differs significantly from the minimizer of the objective function.

As already anticipated, the trust region approach can keep this problem under control by restricting the search for the minimizer of the model function within the

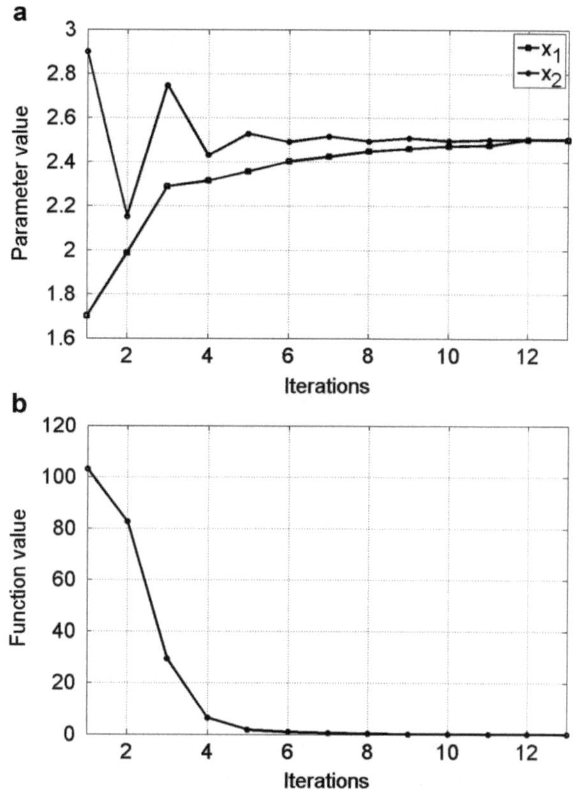

Fig. 2.18 Results of optimization with Cauchy point trust region

trust region zone. The result of the optimization by trust region is visualized in Fig. 2.18.

It may be observed from Fig. 2.18b that the trust region algorithm provided a monotone decrease of the objective function throughout the whole optimization. The implemented strategy for reduction or increase of trust region radius based on the ratio (2.22) turns out to be effective as there was no step rejection in the optimization. In this case starting value of radius was taken to be equal to 0.8. This value was reduced in second iteration and afterwards, as the algorithm started to approach the solution in the fourth iteration it started to be increased in each subsequent iteration.

Previous example showed that the Cauchy point trust region is effective in solving minimization problems in the situations where line search with Newton direction may fail. Trust region has the global convergence property even when the sub-problem is solved only approximately since the Cauchy point doesn't have to be an absolute minimizer of the model function within the trust region.

Since the Cauchy point provides a sufficient reduction of the model function within trust region, and we showed that this is enough for the global convergence, the logical question that arises is why to use any other strategy to solve the sub-problem? The answer is that, if we are using at every iteration Cauchy point we are practically implementing the steepest descend approach on slightly more effective way since the use of model function allowed us to find the minimizer along steepest descend direction more accurately. But as we saw in previous examples the efficiency of steepest descent usually is not connected to the accuracy of finding the minimizer along this direction (e.g., example given in Fig. 2.9). Cauchy point does not depend on the Hessian matrix and it uses it only to compute the value of model function along the steepest descend direction. Rapid convergence can be expected only if Hessian matrix plays an important role in determining also the direction (like in Newton direction line search). There are many trust region algorithms that compute the Cauchy point and then try to improve it. One of them is a so-called *dog-leg* method originally proposed by Powell [6].

2.3.2 Dog-Leg Trust Region

We already saw that in the cases when model function is a good approximation of the objective function and when Hessian is positive definite Newton direction provides quadratic convergence (e.g. result given in Table 2.2). We also saw that similar convergence can be achieved also by modifying Hessian matrix when it is not positive-definite and using Newton or modified Newton direction, where needed, within line search algorithm (result visualizes in Fig. 2.14).

To incorporate fast quadratic convergence rate offered by full Newton step and global convergence feature of steepest descent into a single trust region algorithm an approach called dog-leg method for solving the sub-problem can be used. This method finds a compromise between steepest descent step and Newton's step based on the size of the trust region.

Let $\mathbf{x}_{k+1}^{SD}$ and $\mathbf{x}_{k+1}^{N}$ be steepest descend and Newton step respectively. If the Hessian matrix is positive definite then Newton step is actual minimizer of model function. If this point lays inside the trust region than it should be taken as the solution of the sub-problem. Otherwise, the piecewise linear curve defined by the line segments joining $\mathbf{x}_k$ to $\mathbf{x}_{k+1}^{SD}$ and $\mathbf{x}_{k+1}^{N}$ called *dog-leg trajectory* is taken (Fig. 2.19). The point in which this trajectory intersects trust region, $\mathbf{x}_{k+1}^{DL}$ is taken as the minimizer of the sub-problem.

The dog-leg trajectory can be implemented quit easy. In the first step using the gradient and Hessian matrix a steepest descent direction $\mathbf{p}_k^{SD}$and Newton direction

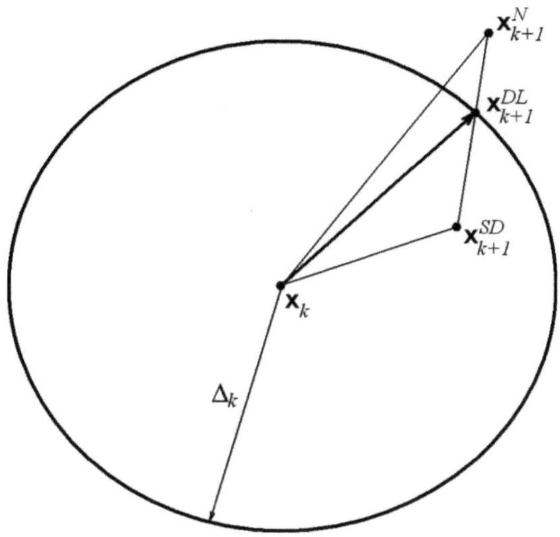

Fig. 2.19 Dog-leg trajectory

$\mathbf{p}_k^N$ are computed. For the case when Newton step lays outside the trust region, an intersection point that corresponds to dog-leg step is found solving the following equation

$$\left\| \mathbf{p}_k^{SD} + \alpha\left(\mathbf{p}_k^N - \mathbf{p}_k^{SD}\right) \right\| = \Delta_k \tag{2.23}$$

This is an effective strategy since both the Cauchy point (guarantying the global convergence) and the full Newton step (ensuring the possibility to have faster rate of convergence) are incorporated into the possible step. It is obvious that the solution will be closer to the Cauchy point for small trust regions while for large enough trust region it will be reduced to Newton's method.

The following listing shows a MATLAB code with the implementation of dog-leg strategy.

```
*****************************************************************
% This is a Trust region algorithm
% A dog-leg approach for sub-problem
Clear

% Setting the options
minchg=1e-4; % Minimum change in parameters
MAXIT=30; % Maximum allowed number of iterations
guess=[1.7;2.9]; % Initial guess of parameters
pert=1e-6; % Perturbation for the first derivatives
res=10;
TRrad=1; % Initial trust region radius
HessMod=0; % Indication for Hessian modification

% Optimization cycle
iter=0;
while res>1e-6
   eV=funLSQ1(guess);
   e0=0.5*eV'*eV;
   itiner(iter+1,1:2)=guess';
   itiner(iter+1,3)=e0;
   iter=iter+1;
   for i=1:size(guess,1)
       guessp=guess;
       guessp(i)=guessp(i)+pert;
       eV=funLSQ1(guessp);
       e1=0.5*eV'*eV;
       grad(i,1)=(e1-e0)/pert;
   end
   % Computing Hessian matrix
   HESS=comhess(@funLSQ1,guess,pert);
   % Ensuring that Hessian is positive-definite
   if HessMod==1
     lambdas=eigs(HESS);
     if lambdas(1)>0 && lambdas(2)>0
         hessmod=0;
     else
         coeff=mean(abs(lambdas));
         posdef=0;
         while posdef<1
             HESSm=HESS+coeff*eye(2);
             hessmod=1; % Indication of modified HESSIAN
             lmb=eigs(HESSm);
             if lmb(1)>0 && lmb(2)>0
                 posdef=1;
             else
                 coeff=coeff*1.5;
             end
         end
     end
```

```
    else
        hessmod=0;
    end
    stpdsc=-grad/norm(grad);
    if hessmod==0
        newton=-inv(HESS)*grad;
    else
        newton=-inv(HESSm)*grad;
    end
    accepted=0;
    while accepted<1
    if norm(newton)<TRrad
        pDL=newton; % Dog Leg step
    else
        % Finding the Cauchy point
        pc=cauchypnt(e0,stpdsc,grad,HESS,TRrad);
        % Finding minimizer within trust region (Dog Leg step)
        diff=newton-pc;
        dimV=size(newton,1);
        cf=[0,0,-TRrad^2];
        for ii=1:dimV
            cf(1)=cf(1)+diff(ii)^2;
            cf(2)=cf(2)+2*pc(ii)*diff(ii);
            cf(3)=cf(3)+pc(ii)^2;
        end
        alfa=max(roots(cf)); % Taking the positive root
        pDL=pc+alfa*diff;
    end
        predred=-(pDL'*grad+0.5*pDL'*HESS*pDL);
        guess1=guess+pDL; % Next iterate
        eV=funLSQ1(guess1);
        e1tr=0.5*eV'*eV;
        realred=e0-e1tr;
        ratio=realred/predred;
        if ratio<0
            TRrad=TRrad/1.2;
        else
            accepted=1;
            if ratio<0.2
                TRrad=TRrad/1.2;
            end
            if ratio>0.6
                TRrad=TRrad*1.2;
            end
        end
    end
    itiner(iter+1,1:2)=guess1';
    itiner(iter+1,3)=e1tr;
    guess=guess1;
    res=e1tr;
    if iter>MAXIT % Terminated after the iteration reach MAXIT
        res=0;
    end
```

```
   % Checking convergence options
   if abs(itiner(iter+1,1)-itiner(iter,1))<minchg ||
abs(itiner(iter+1,2)-itiner(iter,2))<minchg;
       res=0;
   end
end
************************************************************
```

As addition to this MATLAB routine, functions `cauchypnt`, `funLSQ1` and `comhess` with listings given previously need to be used.

The results of the optimization using dog-leg trust region, starting from the same initialization point like for the Cauchy point are visualized in Fig. 2.20. The algorithm proved to be more effective than Cauchy point and it reaches the convergence after seven iterations. It is interesting to note that, dog-leg shows more effective behavior particularly in the zone of the solution as then it uses the full Newton step and therefore is not experiencing the slow approaching to the solution characteristic for steepest descend direction (for example this can be notice in Fig. 2.18 from iteration 8 to iteration 13).

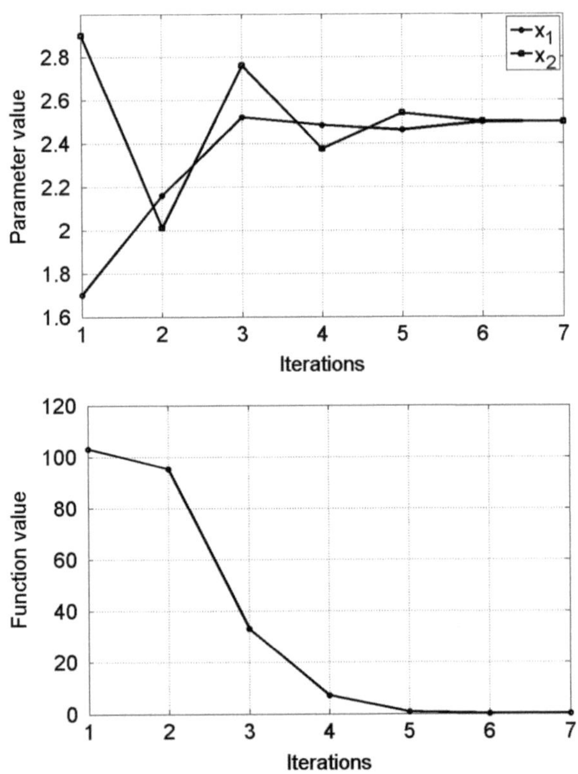

Fig. 2.20 Results of optimization with dog-leg trust region

The implementation given in the above listing takes also into account possibility of modifying Hessian matrix if it is not positive definite on the same way like previously presented for the Newton line search direction. It should be mentioned however that this might not be necessary for the trust region strategy as it introduces its own modification by constraining the minimization of sub-problem. In fact, if we use the same optimization problem starting from parameter set $\mathbf{x} = [1.5, 2.9]^{\mathrm{T}}$, the one which with the Newton direction line search faced the problem of non-positive definite Hessian (Fig. 2.13), with dog-leg trust region it works even without Hessian modification (parameter `HessMod` set to 0) with initial trust region radius set to 1. Since in all steps the dog-leg algorithm makes a combination between steepest descent direction and the Newton direction, this already regularizes the step to be descending in each iteration even without Hessian modification.

The dog-leg method can be made slightly more sophisticated by widening the search for $\mathbf{p}$ to the entire two-dimensional subspace spanned by gradient direction and Newton direction. This approach is described in what follows.

2.3.3 Two-Dimensional Subspace Minimization

By writing the unknown direction as a linear combination of Newton and steepest descend direction, the sub-problem will obtain the following form

$$\begin{aligned}\min m_k(\mathbf{p}_k) = & f(\mathbf{x}_k) + \left[\alpha_1 \mathbf{p}^{SD} + \alpha_2 \mathbf{p}^{N}\right]^T \nabla f(\mathbf{x}_k) \\ & + \frac{1}{2}\left[\alpha_1 \mathbf{p}^{SD} + \alpha_2 \mathbf{p}^{N}\right]^T \cdot \nabla^2 f(\mathbf{x}_k) \cdot \left[\alpha_1 \mathbf{p}^{SD} + \alpha_2 \mathbf{p}^{N}\right]\end{aligned} \tag{2.24}$$

under the constrain

$$\left\|\alpha_1 \mathbf{p}_k^{SD} + \alpha_2 \mathbf{p}_k^{N}\right\| \leq \Delta_k \tag{2.25}$$

The problem now becomes two dimensional and it is solved for the unknown coefficients α_1 and α_2.

This represents a more general version of dog-leg method, since the solution provided by the dog-leg approach is a part of this 2D sub-space minimization problem.

The reduction of model function achieved by the two-dimensional sub-space minimization strategy is often very close to the reduction achieved by exact solution of (2.21). Since Cauchy point is a feasible solution, the reduction achieved will be at least as the one obtained by Cauchy point, resulting in global convergence. On the other hand, wider search for the minimizer with respect to dog-leg method often provides better reduction.

The solution of two-dimensional sub-problem (2.24) and (2.25) represents a constrained minimization problem. We should recall that within dog-leg approach, the sub-problem is not solved like constrained minimization problem. Even though the solution satisfies the trust region constrain, the problem is solved in two steps, using geometrical approach in which it is reduced to Eq. 2.23. Here on the other hand, we will solve the sub-problem as a classical minimization problem with inequality constrain.

This minimization problem, defend by (2.24), with one inequality constrain given by (2.25) can be solved with Lagrange multipliers technique. As a first step we should write the minimization function together with its constrain in the following form

$$\ell(\mathbf{p}_k, \lambda_1) = m_k(\mathbf{p}_k) - \lambda_1 \cdot c_1(\mathbf{p}_k) \tag{2.26}$$

Last equation represents a so-called Lagrangian function, where

$$c_1(\mathbf{p}_k) = \Delta_k - \left\|\alpha_1 \mathbf{p}_k^{SD} + \alpha_2 \mathbf{p}_k^N\right\| \geq 0 \tag{2.27}$$

is called a constrain function, and scalar λ_1 represents a Lagrange multiplier. The solution of this constrained minimization problem is found as a stationary point of the Lagrangian function. Therefore, if we write the Lagrangian function in the general form for an arbitrary number of constrain functions it will adopt the following form

$$\ell(\mathbf{p}_k, \lambda_i) = m(\mathbf{p}_k) - \sum_{i=1}^{N} \lambda_i c_i(\mathbf{p}_k) \tag{2.28}$$

A general condition that $\mathbf{p}_k^*$ needs to satisfy in order to represent a local solution of this minimization problem with inequality constrains is defined as follows

$$\nabla_{\mathbf{p}_k} \ell(\mathbf{p}_k^*, \lambda^*) = 0 \tag{2.29a}$$

$$c_i(\mathbf{p}_k^*) \geq 0, \quad i = 1, ..N \tag{2.29b}$$

$$\lambda_i^* \geq 0, \quad i = 1, ...N \tag{2.29c}$$

$$\lambda_i^* c_i(\mathbf{p}_k^*) = 0 \tag{2.29d}$$

This set of conditions is also known as *Karush-Kuhn-Tucker condition*, or *KKT condition*. This condition practically states, that the solution of the minimization problem should satisfy stationary condition taking the partial derivatives with respect to all the components of vector $\mathbf{p}_k$ and all Lagrange multipliers. Derivatives of Lagrangian with respect to components of vector $\mathbf{p}_k$ results in algebraic equation (2.29a), while partial derivatives of Lagrangian with respect to multipliers are resulting in constrain inequalities (2.29b). Equation 2.29c state that only non-negative Lagrange multipliers should be taken into account, while (2.29d) are complementarity conditions which imply that, either i^{th} constrain is active or otherwise λ_i^* should be zero. The Lagrange multipliers corresponding to inactive inequalities are zeros, and they can be omitted when writing the general problem with (2.28).

A practical procedure of applying the KKT condition to the present problem defined by (2.26) and (2.27) can proceed as follows. In the first step one should first check if there is a minimizer of the model function m_k within the trust region. It practically means that one should find the unconstrained minimizer, which is found from the condition $\nabla_{\mathbf{p}_k} m_k = 0$, namely finding the first derivatives only of model function with respect to unknown direction. The solution of this problem is obviously Newton step. If Hessian is positive definite, it means that the Newton step is minimizer of model function and if it is inside the TR than it should be taken as a result of minimization problem and the constrain (2.27) is not active, so the conditions (2.29a - 2.29d) are actually reduced to the unconstrained problem solution. It is consistent with the KKT condition since, by having the constrain not active, imposes zero value to Lagrange multiplier which automatically means that the compatibility condition are satisfied.

If Hessian is not positive definite it means that the Newton direction is not pointing to the minimum and it is not the solution of the minimization problem. In this case the minimizer of the constrained problem should be searched on the boundary, meaning that the constrain is active. In this case the problem is solved as minimization with equality constrain, by taking all the partial derivatives with respect to vector $\mathbf{p}_k$ components and Lagrange multipliers and solving the obtained system of algebraic equations for non-negative Lagrange multipliers and vector components $\mathbf{p}_k$ simultaneously.

In the case where Newton direction doesn't exist, the constrain is also active and the problem is solved in the same way as mentioned above.

To illustrate this last case let us assume that the function we would like to minimize has the following form

$$f(\mathbf{x}) = x_1 + x_2 \tag{2.30}$$

Les us assume that we are at the point defined by $\mathbf{x}_k = [1,1]^{\mathrm{T}}$, and that the trust region radius is equal to 1. The minimization sub-problem is therefore defined by

$$\min m(\mathbf{p}_k) = 2 + p_1 + p_2, \quad c_1(\mathbf{p}) = 1 - p_1^2 - p_2^2 \leq 0 \tag{2.31}$$

Obviously, since in this case Hessian matrix is zero, the Newton direction doesn't exist. In fact, the objective function is monotonely decreasing and it doesn't have any minimum in the unconstrained domain. The problem therefore should be solved with the active constrain, and the minimizer should be found on the boundary, which in this case is a circle. Applying KKT condition, (2.29a) and (2.29b) are resulting in the following algebraic equations

$$1 + 2\lambda_1 p_1 = 0$$

$$1 + 2\lambda_1 p_2 = 0$$

$$p_1^2 + p_2^2 = 1 \tag{2.32}$$

Expressing from the first two equations p_1 and p_2 as functions of λ_1 and substituting to the third one, it results in a quadratic equation that should be solved for the positive values of λ_1. It is easy to see that the solution of this quadratic equation gives

$$\lambda_1 = \pm\frac{1}{\sqrt{2}} \tag{2.33}$$

Taking just the positive value in order to satisfy (2.29c) the resulting minimizer of the constrained problem is equal to

$$\mathbf{p}_k = \left[-\frac{\sqrt{2}}{2}, -\frac{\sqrt{2}}{2}\right] \tag{2.34}$$

that is the point on the circle with unit radius with the center in $[1,1]^{\mathrm{T}}$ for which the function (2.30) has its minimum value.From the above discussion it can be summarized that the minimizer of the constrained sub-problem that arises within trust region algorithm is taken to be the Newton step in the case when Hessian is positive definite, and when Newton step lays within the trust region. If this is not the case, it means that, either the model function doesn't have the minimum (it monotonely decreases) or the minimizer is found outside of the trust region. In both of these cases the solution should be found on the boundary itself. As demonstrated in previous simple example this will lead us to the system of algebraic equations, like those given in (2.32), where from the first set of equations all the components of unknown direction should be expressed as a function of λ_1, which after introducing to the last equation will result in a polynomial equation, that is further solved for the positive roots.

Two-dimensional sub-problem defined by (2.24) and (2.25) obviously represent a special case of the presented strategy, where the unknown direction is uniquely defined with two scalars, α_1 and α_2, that are used as coefficients of linear combination of two known directions (steepest descend and Newton direction). Using the same approach as previously summarized, we should first compute steepest descend and Newton direction using (2.9) and (2.15) respectively. If the Hessian is positive definite, it means that the Newton step is the minimizer of the model function and then, we should check whether it lays inside the trust region. If this is the case, it is taken as the solution of the sub-problem. In the opposite case, we should proceed with the solution of constrained problem. For general multidimensional case (when the number of parameters is N), Lagrangian will take the following form

$$\ell(\mathbf{p}_k, \lambda_i) = f(\mathbf{x}_k) + \left[\alpha_1 p_i^{SD} + \alpha_2 p_i^N\right]^T \cdot [g_i] \cdot + \frac{1}{2}\left[\alpha_1 p_i^{SD} + \alpha_2 p_i^N\right]^T \cdot \left[h_{ij}\right] \cdot \left[\alpha_1 p_i^{SD} + \alpha_2 p_i^N\right]^T + \lambda_1\left(\Delta^2 - \sum_{i=1}^{N}\left(\alpha_1 p_i^{SD} + \alpha_2 p_i^N\right)\right) \tag{2.35}$$

whew $[g_i]$ is gradient vector, $[h_{ij}]$ is Hessian matrix and $[p_i^{SD}]$ and $[p_i^N]$ are steepest descend and Newton direction vectors respectively. Writing the first derivatives with respect to α_1, α_2 and λ_1 will lead us to the following system of algebraic equations

$$C_{11}\alpha_1 + C_{12}\alpha_2 + 2L_{11}\alpha_1\lambda_1 + L_{12}\alpha_2\lambda_1 = F_1 \tag{2.36a}$$

$$C_{21}\alpha_1 + C_{22}\alpha_2 + L_{21}\alpha_1\lambda_1 + 2L_{22}\alpha_2\lambda_1 = F_1 \tag{2.36b}$$

$$L_{11}\alpha_1^2 + L_{12}\alpha_1\alpha_2 + L_{22}\alpha_2^2 = \Delta^2 \tag{2.36c}$$

where the coefficients are given by

$$C_{11} = \sum_{i=1}^{N} p_i^{SD} \sum_{j=1}^{N} h_{ij} p_j^{SD} \tag{2.37a}$$

$$C_{12} = C_{21} = \frac{1}{2}\left(\sum_{i=1}^{N} p_i^{SD} \sum_{j=1}^{N} h_{ij} p_j^{N} + \sum_{i=1}^{N} p_i^{N} \sum_{j=1}^{N} h_{ij} p_j^{SD}\right) \tag{2.37b}$$

$$C_{22} = \sum_{i=1}^{N} p_i^{N} \sum_{j=1}^{N} h_{ij} p_j^{N} \tag{2.37c}$$

$$F_1 = -\sum_{i=1}^{N} g_i p_i^{SD} \tag{2.37d}$$

$$F_2 = -\sum_{i=1}^{N} g_i p_i^{N} \tag{2.37e}$$

$$L_{11} = \sum_{i=1}^{N} \left(p_i^{SD}\right)^2 \tag{2.37f}$$

$$L_{12} = L_{21} = 2\sum_{i=1}^{N} p_i^{SD} \cdot p_i^N \tag{2.37g}$$

$$L_{22} = \sum_{i=1}^{N} \left(p_i^N\right)^2 \tag{2.37h}$$

Further, the procedure is the same as previously shown. From the first two equations α_1 and α_2 are expressed as functions of λ_1, which after being introduced to the third one results in polynomial equation that is solved for positive values of λ_1. Since in general, depending on the values of Hessian matrix and gradient this polynomial can be of higher order, it may have larger number of positive real roots. However, once they are identified it is computationally inexpensive to check which value corresponds to α_1 and α_2 that minimize the objective function.

The reduction of the model function achieved by two-dimensional subspace minimization is often very close to the exact solution of (2.21). The solution of dog-leg method is obviously included in two-dimensional subspace method so at the limit case the performance should match those achieved by the former approach. Usually the advantage of two-dimensional sub-space approach is evidenced in larger reduction of model function in each step. More details of this approach can be found in [7] and [8].

Expressing the polynomial from the system (2.36a - 236c) only in terms of Lagrange multipliers results in rather complicated equation and the implementation is relatively long. In order to shown the abovementioned benefits that are achieved by this approach we can focus on simpler case where the number of parameters to identify is equal to 2. In this case, by considering derivatives directly with respect to components of vector $\mathbf{p}_k$, the general form of the solution is somewhat simpler, and it leads to the same solution as the one achieved by the two-dimensional sub-problem approach.

Adopting the same notation as in previous case, the model function will take the following form

$$m(\mathbf{p}) = f_k + \begin{bmatrix} p_1 & p_2 \end{bmatrix} \begin{bmatrix} g_1 \\ g_2 \end{bmatrix} + \frac{1}{2} \begin{bmatrix} p_1 & p_2 \end{bmatrix} \begin{bmatrix} h_{11} & h_{12} \\ h_{21} & h_{22} \end{bmatrix} \begin{bmatrix} p_1 \\ p_2 \end{bmatrix} \tag{2.38}$$

with the constrain function given by

$$c(\mathbf{p}) = \Delta^2 - p_1^2 - p_2^2 \geq 0 \tag{2.39}$$

After the matrix multiplication in (2.38) is performed, and considering that Hessian is symmetric, Lagrangian function can be written as follows

$$\ell = f_k + g_1 p_1 + g_2 p_2 + \frac{1}{2} h_{11} p_1^2 + h_{12} p_1 p_2 + \frac{1}{2} h_{22} p_2^2 - \lambda\left(1 - p_1^2 - p_2^2\right) \tag{2.40}$$

Stationary point of Lagrangian results in the following system of algebraic equations

$$\frac{\partial \ell}{\partial p_1} = g_1 + h_{11}p_1 + h_{12}p_2 + 2\lambda p_1 = 0 \tag{2.41a}$$

$$\frac{\partial \ell}{\partial p_2} = g_2 + h_{12}p_1 + h_{22}p_2 + 2\lambda p_2 = 0 \tag{2.41b}$$

$$\frac{\partial \ell}{\partial \lambda} \Rightarrow p_1^2 + p_2^2 = \Delta^2 \tag{2.41c}$$

From (2.41b) we can express p_2 as a function of p_1 and λ resulting in

$$p_2 = -\frac{h_{12}p_1 + g_2}{2\lambda + h_{22}} \tag{2.42}$$

Combining (2.42) with (2.41a) we can obtain expression of p_1 only as a function of λ

$$p_1 = -\frac{2g_1\lambda + g_1h_{22} - h_{12}g_2}{4\lambda^2 + (2h_{11} + 2h_{22})\lambda + h_{11}h_{22} - h_{12}^2} \tag{2.43}$$

Finally, combining the last two equations also p_2 component can be expressed only as a function of λ resulting in

$$p_2 = \frac{-4g_2\lambda^2 + (2g_1h_{12} - 2h_{11}g_2 - 2h_{22}g_2)\lambda + g_1h_{12}h_{22} - g_2h_{11}h_{22}}{(2\lambda + h_{22})\left[4\lambda^2 + (2h_{11} + 2h_{22})\lambda + h_{11}h_{22} - h_{12}^2\right]} \tag{2.44}$$

Equations 2.43 and 2.44 are substituted back in (2.41c) to obtain polynomial of λ.

In a view of easier implementation we can group the coefficients in the following way. Considering that both p_1 and p_2 are polynomial functions of λ we can write the (2.41c) in the following way

$$\frac{(b_1\lambda + a_1)^2}{\left(c_2\lambda^2 + b_2\lambda + a_2\right)^2} + \frac{\left(c_3\lambda^2 + b_3\lambda + a_3\right)^2}{(b_4\lambda + a_4)^2\left(c_2\lambda^2 + b_2\lambda + a_2\right)^2} = \Delta^2 \tag{2.45}$$

where the introduced coefficients are computed using the following equations

$$b_1 = 2g_1 \tag{2.46a}$$

$$a_1 = g_1h_{22} - h_{12}g_2 \tag{2.46b}$$

$$c_2 = 4 \tag{2.46c}$$

$$b_2 = 2h_{11} + 2h_{22} \tag{2.46d}$$

$$a_2 = h_{11}h_{22} - h_{12}^2 \tag{2.46e}$$

$$c_3 = -4g_2 \tag{2.46f}$$

$$b_3 = 2g_1h_{11} - 2h_{11}g_2 - 2h_{22}g_2 \tag{2.46g}$$

$$a_3 = g_1h_{11}h_{22} - g_2h_{11}h_{22} \tag{2.46h}$$

$$b_4 = 2 \tag{2.46i}$$

$$a_4 = h_{22} \tag{2.46j}$$

After multiplication, 2.45 is transformed into a single polynomial equation of sixth order

$$\begin{aligned}
&-\Delta b_4^2c_2^2\lambda^6 - \left(2\Delta^2 c_2b_2b_4^2 + 2\Delta^2 a_4b_4c_2^2\right)\lambda^5 \\
&+\left(b_1^2b_4^2 + c_3^2 - \Delta^2 b_2^2b_4^2 - 2\Delta^2 a_2c_2b_4^2 - 4\Delta^2 c_2b_2a_4b_4 - \Delta^2 a_4^2c_2^2\right)\lambda^4 \\
&+(2b_1^2b_4a_4 + 2b_1a_1b_4^2 + 2c_3b_3 - 2\Delta^2 a_2b_2b_4^2 - 2\Delta^2 a^4b_4b_2^2 - 4\Delta^2 a_2c_2a_4b_4 \\
&-2\Delta^2 a_4^2c_2b_2)\lambda^3 + (b_1^2a_4^2 + 4b_1a_1b_4a_4 + a_1^2b_4^2 + b_3^2 + 2c_3a_3 - \Delta^2 a_2^2b_4^2 - 4\Delta^2 a_2b_2a_4b_4 \\
&-\Delta^2 a_4^2b_2^2 - 2\Delta^2 a_4^2a_2c_2)\lambda^2 + (2b_1a_1a_4^2 + 2a_1^2b_4a_4 + 2a_3b_3 - 2\Delta^2 a_4b_4a_2^2 \\
&-2\Delta^2 a_4^2a_2b_2)\lambda + a_1^2a_4^2 + a_3^2 - \Delta^2 a_4^2a_2^2 = 0
\end{aligned} \tag{2.47}$$

The resulting polynomial is sixth order, but in general not all the roots are real positive numbers. Therefore, in the practical implementation, we will first compute the values of coefficients using Eq. 2.46, and then we will compute roots of polynomial (2.47). Afterwards, only those that are positive real numbers will be considered to compute resulting direction components p_1 and p_2 using (2.43) and (2.44) respectively. In most of the cases there will be more than one pair of components corresponding to real positive Lagrange multipliers. However, once they are computed, by using Eq. 2.38. it is computationally inexpensive to verify which one of the computed directions minimizes the model function.

The implementation of presented procedure is given in the listing below.

```
*************************************************************
% This is a Trust region algorithm
% with minimization for components of vector p

clear
% Setting the options
minchg=1e-4; % Minimum change in parameters
MAXIT=30; % Maximum allowed number of iterations
guess=[1.7;2.9];
pert=1e-6; % Perturbation for the first derivatives
res=10;
TRrad=0.8;
HessMod=1; % Indication for Hessian modification

% Optimization cycle
iter=0;
while res>1e-6
eV=funLSQ1(guess);
e0=0.5*eV'*eV;
itiner(iter+1,1:2)=guess';
itiner(iter+1,3)=e0;
iter=iter+1;
for i=1:size(guess,1)
    guessp=guess;
    guessp(i)=guessp(i)+pert;
    eV=funLSQ1(guessp);
    e1=0.5*eV'*eV;
    grad(i,1)=(e1-e0)/pert;
end
% Computing Hessian matrix
HESS=comhess(@funLSQ1,guess,pert);
% Ensuring that Hessian is positive-definite
if HessMod==1
lambdas=eigs(HESS);
if lambdas(1)>0 && lambdas(2)>0
    hessmod=0;
else
    coeff=mean(abs(lambdas));
    posdef=0;
    while posdef<1
        HESSm=HESS+coeff*eye(2);
        hessmod=1; % Indication of modified HESSIAN
        lmb=eigs(HESSm);
        if lmb(1)>0 && lmb(2)>0
            posdef=1;
        else
            coeff=coeff*1.5;
        end
    end
end
else
    hessmod=0;
```

```
end
stpdsc=-grad/norm(grad);
if hessmod==0
    newton=-inv(HESS)*grad;
else
    newton=-inv(HESSm)*grad;
end
accepted=0;
while accepted<1
    % Solving for vector p that minimizes the quadratic form
    if norm(newton)<TRrad
        pfnd=newton;
    else
    % Constrain should be active
    % Coefficients for components of vector p
    B1=2*grad(1);
    A1=grad(1)*HESS(2,2)-HESS(1,2)*grad(2);
    C2=4;
    B2=2*HESS(1,1)+2*HESS(2,2);
    A2=HESS(1,1)*HESS(2,2)-HESS(1,2)^2;
    C3=-4*grad(2);
    B3=2*grad(1)*HESS(1,1)-2*HESS(1,1)*grad(2)-
2*HESS(2,2)*grad(2);
    A3=grad(1)*HESS(1,1)*HESS(2,2)-
grad(2)*HESS(1,1)*HESS(2,2);
    B4=2;
    A4=HESS(2,2);
    % Coefficients for polynomial for Lagrange multiplier
    CF(1)=-TRrad^2*B4^2*C2^2;
    CF(2)=-2*TRrad^2*C2*B2*B4^2-2*TRrad^2*A4*B4*C2^2;
    CF(3)=B1^2*B4^2+C3^2-TRrad^2*B2^2*B4^2-
2*TRrad^2*A2*C2*B4^2-4*TRrad^2*C2*B2*A4*B4-TRrad^2*A4^2*C2^2;
    CF(4)=2*B1^2*B4*A4+2*B1*A1*B4^2+2*C3*B3-
2*TRrad^2*A2*B2*B4^2-2*TRrad^2*A4*B4*B2^2-
4*TRrad^2*A2*C2*A4*B4-2*TRrad^2*A4^2*C2*B2;
    CF(5)=B1^2*A4^2+4*B1*A1*B4*A4+A1^2*B4^2+B3^2+2*C3*A3-
TRrad^2*A2^2*B4^2-4*TRrad^2*A2*B2*A4*B4-TRrad^2*A4^2*B2^2-
2*TRrad^2*A4^2*A2*C2;
    CF(6)=2*B1*A1*A4^2+2*A1^2*B4*A4+2*A3*B3-
2*TRrad^2*A4*B4*A2^2-2*TRrad^2*A4^2*A2*B2;
    CF(7)=A1^2*A4^2+A3^2-TRrad^2*A4^2*A2^2;
    LAMBDAS=roots(CF);
    rr=0;
    clear LMB
    for i=1:6
        if isreal(LAMBDAS(i))==1
            rr=rr+1;
            LMB(rr)=LAMBDAS(i);
        end
    end
    % Computing vector p components
    accp=0;
    clear p
```

```
    for i=1:rr
        accp=accp+1;
        p(1,accp)=-(B1*LMB(i)+A1)/(C2*LMB(i)^2+B2*LMB(i)+A2);
        p(2,accp)=(C3*LMB(i)^2+B3*LMB(i)+A3)/
((B4*LMB(i)+A4)*(C2*LMB(i)^2+B2*LMB(i)+A2));
        if norm(p(:,accp))>(TRrad+0.01)
            accp=accp-1;
        end
    end
    % Checking which solution gives minimum of model function
    minMOD=e0+(p(:,1)'*grad+0.5*p(:,1)'*HESS*p(:,1));
    pfnd=p(:,1);
    for i=2:accp
        modP=e0+(p(:,i)'*grad+0.5*p(:,i)'*HESS*p(:,i));
        if modP<minMOD
            pfnd=p(:,i);
            minMOd=modP;
        end
    end
    end
    predred=-(pfnd'*grad+0.5*pfnd'*HESS*pfnd);
    guess1=guess+pfnd; % Next iterate
    eV=funLSQ1(guess1);
    e1tr=0.5*eV'*eV;
    realred=e0-e1tr;
    ratio=realred/predred;
    if ratio<0
        TRrad=TRrad/1.2;
    else
        accepted=1;
        if ratio<0.2
            TRrad=TRrad/1.2;
        end
        if ratio>0.6
            TRrad=TRrad*1.2;
        end
    end
end
itiner(iter+1,1:2)=guess1';
itiner(iter+1,3)=e1tr;
guess=guess1;
res=e1tr;
if iter>MAXIT % Terminated after the iteration reaches MAXIT
    res=0;
end
% Checking convergence options
if abs(itiner(iter+1,1)-itiner(iter,1))<minchg ||
abs(itiner(iter+1,2)-itiner(iter,2))<minchg;
    res=0;
end
end
************************************************************
```

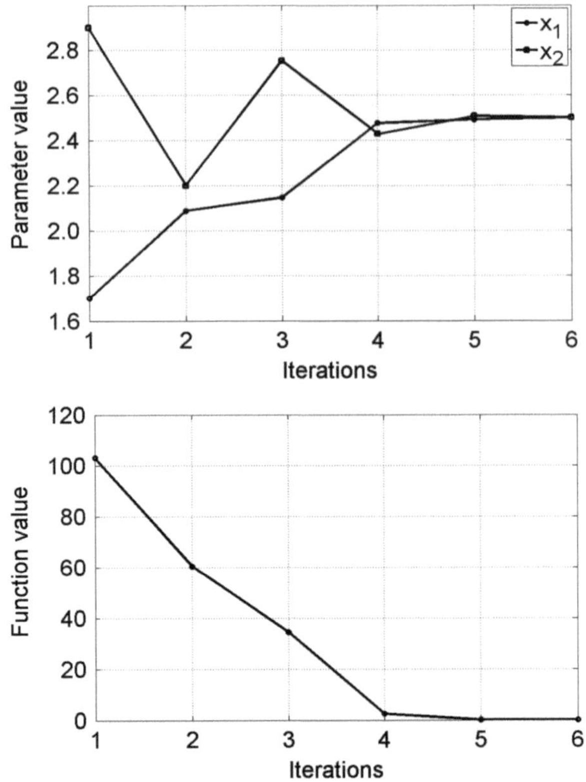

Fig. 2.21 Results of optimization with direct minimization on direction vector components

This algorithm is used to solve the same optimization problem as the one already solved with dog-leg approach (visualized in Fig. 2.20). The results of the optimization are visualized in the same manner as previously in Fig. 2.21. It may be observed that the latter algorithm turned out to be more effective, as expected, and the optimization terminated after six iterations (one less with respect to dog-leg approach).

Comparing the two figures it may be observed that the gain is not significant. In fact, the only thing in which the second approach is more effective is the minimization of the model function under each step. However, also the dog-leg approach is not far from the solution and so the steps are not much different.

This difference can be illustrated if we compare what are the values of model function in the resulting step within each iteration for both algorithms. However, this comparison is fair to make only at the beginning of the optimization if both algorithms are starting from the same point, since, as they will produce different steps in every second iteration the two algorithms will not be at the same point and therefore will not have the same model function to minimize. These differences are given in Table 2.3, referring to the first iteration of three different initialization points.

Table 2.3 Values of model function at the beginning and the end of the first iterations obtained by two different optimization algorithms referring to three different initializations

	Dog–leg		Minimization with respect to **p**	
Initialization	m_k at the beginning	m_k at the end of step	m_k at the beginning	m_k at the end of step
[1.7, 2.9]	103.1	−35.3	103.1	−39.9
[1.3, 2.8]	133.4	16.8	133.4	14.4
[1.4, 2.95]	155.8	−33.8	155.8	−33.8

It may be observed from the table that the second approach indeed is obtaining lower values of model function in most of the cases. Since the solution by dog-leg strategy is part of the second approach the latter should perform at least like the former one, which was the case in the third initialization reported in the table, where the reduction of model function achieved by both algorithms was the same. This small advantage of more precise minimization of the model function in some of the cases may result in less iterations. Example of this is the third initialization given in Table 2.3, that converged by dog-leg approach after 8 iterations with respect to 7 obtained by the second approach.

However, in most of the cases the differences are not so large, and also the dog-leg approach manages to minimize the model function almost to the same extent as the minimization with respect to direction vector, so it can be used with almost the same efficiency.

The optimization algorithms presented up to now are some times in the literature found under the name of *derivative based algorithms*, as they rely on the computation of derivatives. This circumstance makes them rather ineffective if the objective function has a large number of local minima, since they are identifying as a solution any mathematical minimum (i.e. first derivative equal to zero and second derivative larger than zero). As a remedy, in these situations Genetic Algorithms can be used as an alternative.

2.4 Genetic Algorithms

The objective functions arising in inverse analyses of structural problems sometimes can be complicated and extremely non-convex. As previously mentioned, in these situations algorithms relying on classical mathematical theory of optimization are behaving rather poor, since they can identify as a solution of the problem also the local minimum.

Using any of derivative based algorithms in order to minimize the function of the type visualized in Fig. 2.22, depending on the initialization point it may occur that the optimization terminates after the point B is reached. Since in point B the first derivative is equal to zero and the second derivative is larger than zero this point represents mathematical minimum, and the algorithm doesn't have information that

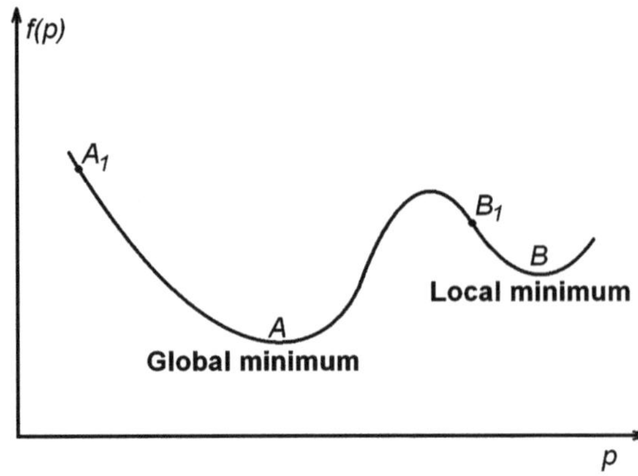

Fig. 2.22 Objective function with one local and one global minimum

somewhere within the range of interest there might be a point with lower value of the objective function. Since derivative based algorithms are stopping when the minimum of the objective function within a certain zone is found, the only way to confirm that it also represents a global minimum of the function is to perform the minimization procedure couple of times starting from different initialization points. If for a given problem it turns out that the result of minimization doesn't depend on initialization point then there is a large probability that the identified minimum is an absolute minimum of the function. For example, minimizing the objective function of the type visualized in Fig. 2.22, by the use of any of derivative based algorithms, starting from point B_1, the algorithm would most likely terminate by identifying point B as a minimum. On the other hand, repeating the same procedure but starting from a different point, say A_1, the algorithm would probably identify point A as a solution. By comparing the values of the objective functions between the two points it's easy to verify which one of the two is the solution of the problem. In fact, common practice in inverse analyses when derivative based algorithms are used is to perform minimization couple of times and to take as the solution parameter set to which procedure converged three or four times starting from different initialization points.

Using this strategy, problems like the one visualized in Fig. 2.22 can still be relatively effectively solved by derivative based algorithms, as the number of local minima is small. If the objective function is characterized by a presence of large number of local minima, then the use of derivative based algorithms is not effective as the solution becomes extremely dependant on the initialization point, and therefore it is difficult to identify what is the absolute minimum of the function. In these situations it is more appropriate to use Genetic Algorithms to solve the minimization problem.

The Genetic Algorithms (GA) represent a methodology for solving both constrained and unconstrained optimization problems. They belong to a so-called *soft-computing* family as they are not solving in mathematical sense the minimization problems. The approach adopted by GA consists in repeated modification of "population" of individual solutions, based on the concept of "natural" selection. Over successive generations, the population "evolves" toward an optimal solution.

GA can be used in general optimization problems, but their real advantage comes to the play in optimizing discontinuous, non-differentiable, stochastic, or highly nonlinear objective functions.

In order to understand the way GA works let us describe the main concept and notions of GA within the present context. Each population consists of a certain number of individuals. For the given optimization problem, the number of members is fixed and is usually about 50–100. Each member is represented by a set of parameters, which are called *genes* in the jargon of GA. To each of the individuals it is attributed some *score* which is further used as a selection criterion. This score represents a value of *fitness function* which in the present context is the value of the objective function for the parameters that are defining a certain individual.

Optimization by GA starts by a random selection of individuals that are covering some region of interest. The first step consists of computing the values of objective function for each of the individuals. After these computations are performed, all the individuals are sorted according to their value of the objective function. In the subsequent step, a new generation is created by making a use of the existing one.

There are many different types of GA that differ based on a way how they form the new generation. For a more detailed description on GA readers should refer to [9] and [10]. In what follows, some of more frequently used criteria for building a new generation from the current one will be explained.

All of the GA are using some individuals from the current generation, called *parents*, who contribute their *genes* (i.e. parameter values) to their *children.* Algorithms are usually selecting individuals with better values of the fitness function as parents.

An easy to implement scheme of forming the next generation is described in what follows. This scheme works very well with the problems studied in this book. GA of this type turns out to be capable of solving the problems with multiple minima objective functions, as it will be demonstrated on the examples that will follow.

Within this scheme, every new generation is formed by three groups of children:

- Elite children
- Cross-over children
- Mutation children

Elite children represent the individuals with best fitness function within the present generation. These children are directly passed to the next generation without any modification. Of course, the number of elite children represents an optimization parameter to be adjusted but for a successfully working GA it should not be very large. Usually it is about 2–5% of the population size. The existence of elite children is important for the preservation of the individuals with already good value of fitness function

Cross-over children represent individuals that are formed by combining genes of two parents from the current generation. This group is formed by applying an adopted cross-over rule on the selected individuals called parents. As parents usually the individuals from best to intermediate values of the fitness function are selected (i.e. the group takes into account both elite members, but also those with somewhat weaker fitness function result). The number of individuals used as parents represents additional optimization parameter, and it is usually about 50–70% of the population size. Two parents are usually combined to reproduce two children so that the total size of the population would be preserved. After the pairs of parents are randomly coupled from previously selected individuals, cross-over rule is applied in order to form children. The idea of crossing over existing individuals is important as it mixes the parameters of existing individuals. Therefore with this operation by combining already assessed parameters in a different way, possibly improved individuals could be formed, as it may occur for certain individuals that the error on some of the parameters is smaller than for the others.

A very simple cross-over rule is the one that uses two random vectors of the same size as the number of parameters, with only zeroes and ones as entries. Applying this rule to the parent pair, the two children are constructed using these two vectors, where entry 1 means taking the corresponding parameter from the fist parent, while entry 0 means taking it from the second one.

This scheme of crossing-over is illustratively presented in Fig. 2.23. The figure shows optimization problem with 4 parameters. Parameters of one parent are represented by x_i while those of the other one by y_i. The two children in this case are created by the use of two randomly generated cross-over vectors. It should be mentioned that, in the case of smaller number of optimization parameters (e.g,. 3) the scheme can work only with one random extraction and taking the other one as the opposite (e.g,. two cross-over vectors can be [1,1,0] for the first child and [0,0,1] for the second one).

Mutation children are created by introduction a random mutation (i.e. changes) to the parameters applied on the selected individuals from the current population. For this operation usually the worst individuals are taken, as they anyhow represent those that should be wasted, so it is reasonable to try on them a fully random modification as it may produce better individuals. The number of individuals that will be subjected to this operation is what remains when previous two groups are excluded. Apart of the total number of mutated individuals this process is also controlled by the parameter that sets the amount of mutation.

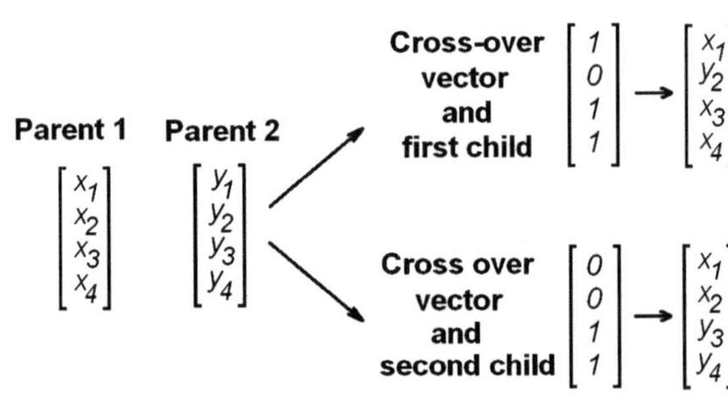

Fig. 2.23 Cross-over scheme with four parameter optimization problem

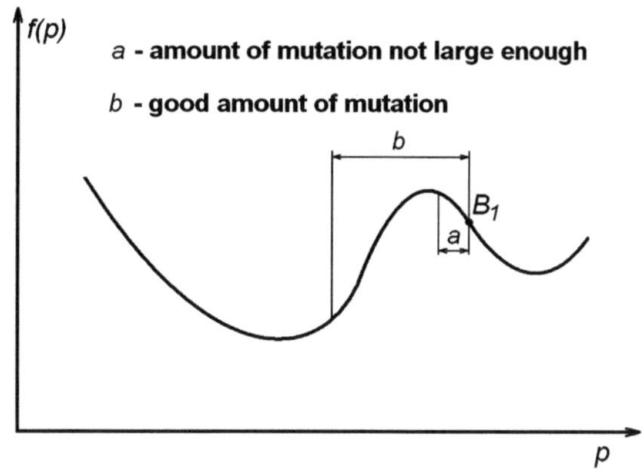

Fig. 2.24 The influence of amount of mutation to the capability of the algorithm to avoid trapping in local minima

Existence of mutation children is very important for the development of GA with the capability to avoid trapping in the local minima. The success of this feature becomes very much dependant on the proper selection of the mutation amount. Figure 2.24 illustrates the influence of this parameter for the one-dimensional case with a local minimum visualized in Fig. 2.22. Obviously by setting a small amount of mutation, the algorithm will not have capability to "jump" out of the local minimum zones.

Applying previously described scheme, next generation is produced that substitutes the current one. This process iteratively continues until the convergence criteria are met.

Genetic algorithms are randomly driven optimization procedure, and in order to have them working properly they need to involve relatively large number of computations. As previously mentioned, in the structural problems here of interest, a successful GA are usually having population of about 50 members, while the number of generations up to the convergence to the global minimum is usually about 100. GA represent a soft computing technique, and so the convergence criteria are also defined in rather loose manner. In most of the cases it is enough to introduce two stopping criteria for a successful implementation of GA. The first one puts an upper bound to the maximum allowed number of generations, while the second one puts a limit to the so-called "stalling" number of generations that represents the number of consecutive generations in which no improvement is obtained in terms of individuals with the lowest value of the objective function. The latter one usually has the value of about 20–40% of the former.

Within presented scheme, the first two rules (i.e. selection for the elite children and cross-over) are contributing to better exploration of the zone that the present generation is currently occupying. On the other hand, mutation rules are contributing to the exploration of the rest of the region for potentially improved individuals.

In order to explain this mechanism let us consider example in which the objective function is of Rastrigin's type defined by the following equation

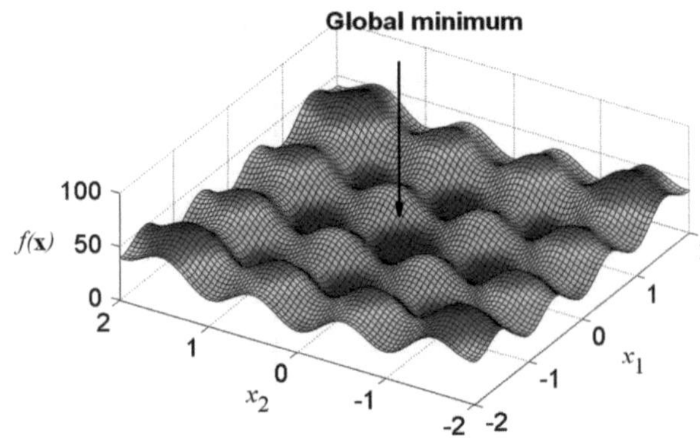

Fig. 2.25 Analytical function with large number of local minima

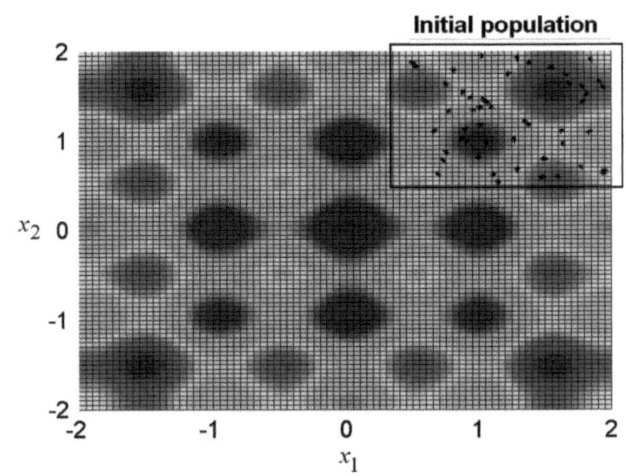

Fig. 2.26 Initial population for GA optimization of analytical function given by Eq. 2.48

$$f(\mathbf{x}) = 20 + 5x_1^2 + 5x_2^2 - 10(\cos 2\pi x_1 + \cos 2\pi x_2) \tag{2.48}$$

The graph of this function for the domain x_1, $x_2 \in [-2,2]$ is visualized in Fig. 2.25. The function is characterized by a large number of local minima, and only one global minimum defined for the coordinate $\mathbf{x} = [0,0]^T$. This function is frequently used to test GA's capability of optimization of extremely non convex functions.

Let us imagine that the optimization by GA starts with initial population which covers one relatively restricted zone of one of the local minima (see Fig. 2.26). As the optimization starts, driven by first two mechanisms (i.e. selection of the elite children and crossing-over of the existing individuals), the individuals will first start to group in the zone of local minima, as for the present range it offers the lowest value of the objective function. On the other hand at each generation there is a set of individuals to whom a random mutation (i.e. perturbation of the parameter values) is attributed, and therefore within couple of generation there is a strong probability that eventually some of them will "jump" out of this zone of local minima.

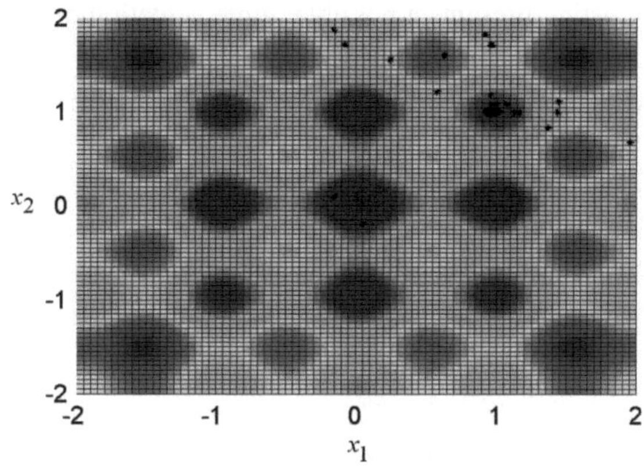

Fig. 2.27 One of the later generations from GA optimization of the analytical function given by Eq. 2.48

Figure 2.27 visualizes some of the later generations of the same optimization. From the figure it may be noticed that even though most of the individuals are grouped in the zone of local minima where the initial population was concentrated, there are two individuals that "jumped" in the zone with smaller values of the objective function. In the following ranking, these will be moved to the elite children and then further by applying cross-over rules they will contribute to the grouping of the individuals in this zone in the following generations.

With this mechanism GA manages to minimize the objective function without mathematically solving the problem. Furthermore, as demonstrated on this example, GA can effectively tackle the problems of multiple local minima as they don't base computation on derivatives and therefore are not terminating the optimization when mathematical minimum is found.

This nice feature of GA however comes for the price of increased computational cost. The total number of objective function evaluations is equal to the product between number of individuals and the total number of generations reached within the optimization procedure. As previously anticipated in structural problems discussed in this book usually these numbers are about 50 members in each population, while the number of generations is usually between 50 and 100. Clearly, the total number of objective function evaluations is at least one order of magnitude larger than what is required when traditional derivative based algorithms are employed. This contributes to the conclusion that it is numerically justifiable to apply GA only in those situations when the objective function is characterize by a large number of local minima, and so the traditional derivative based algorithms would perform poorly.

The following listing presents a possible implementation of discussed GA.

```
************************************************************
clear
clc
% ~~~~~~~~~~~~~~~~~~~~~~~~~~~~~~~~~~~~~~~
% SETTING THE OPTIONS
% ~~~~~~~~~~~~~~~~~~~~~~~~~~~~~~~~~~~~~~~
% Basic parameters
% ~~~~~~~~~~~~~~~~~~~~~~~~~~~~~~~~~~~~~~~
population=100;
elite=2; % Number of individuals that are considered as elite
cross=0.8; % Cross-over ratio
crossP=0.5; % Ratio of individuals to consider as c-o parents
mutR=0.3; % Ratio of individuals to consider for mutation
mutrange=2.5; % Interval of mutation (0.2 means +-0.1)
TOTGEN=200; % Total number of generations
TOTSTALL=80; % Total number of stalling generations
% ~~~~~~~~~~~~~~~~~~~~~~~~~~~~~~~~~~~~~~~
% Computed parameters
% ~~~~~~~~~~~~~~~~~~~~~~~~~~~~~~~~~~~~~~~
% Needed number of the parents to produce next generation
crossN=round(cross*(population-elite));
crsP=round(population*crossP);
mutRN=round(mutR*population);
% Setting the range
npar=2;
range1=[0.5,2];
range2=[0.5,2];
populOld=rand(population,npar);
populOld(:,1)=range1(1)+populOld(:,1)*(range1(2)-range1(1));
populOld(:,2)=range2(1)+populOld(:,2)*(range2(2)-range2(1));
generation=1;
genstl=0;
BEST=1e5;
%~~~~~~~~~~~~~~~~~~~~~~~~~~~~~~~~~~~~~~~~~~~~~~~~~~~~~~~~~~~~~
% beggining of iterations
while generation<TOTGEN
% Calculating fitness values
for i=1:population
    fitness(i,1)=objrastr([populOld(i,1),populOld(i,2)]);
end
% Scoring the result
result(generation,1)=mean(fitness);
result(generation,2)=min(fitness);
% Sorting the population based on value of fitness function
popsort=[fitness,populOld];
popsort=sortrows(popsort,1);
% Checking if the population stalls
if popsort(1,1)<BEST
    BEST=popsort(1,1);
    genstl=0;
else
    genstl=genstl+1;
end
```

```
if genstl>TOTSTALL
    break
end
% Moving the best ones to the elite
eliteKids=popsort(1:elite,2:3);
% Parents for crossover
parX=round(1+(crsP)*rand(crossN,2));
% Individuals for mutation
parMBCK=round(mutRN*rand(population-elite-crossN,1));
parM=population-parMBCK;
% Generating cross-over kids
for i=1:crossN
    first=round(rand(2,1)); % Vector saying which genes will
be taken from the first parent
    second=[1;1]-first;

xKids(i,1)=first(1)*populOld(parX(i,1),1)+second(1)*populOld(p
arX(i,2),1);

xKids(i,2)=first(2)*populOld(parX(i,1),2)+second(2)*populOld(p
arX(i,2),2);
end
% Generating mutation kids
for i=1:population-elite-crossN
    mut=mutrange*rand(2,1)-mutrange/2;
    mKids(i,1)=populOld(parM(i),1)+mut(1);
    mKids(i,2)=populOld(parM(i),2)+mut(2);
end
generation=generation+1;
populOld=[eliteKids;xKids;mKids];
end
% end of iterations
%~~~~~~~~~~~~~~~~~~~~~~~~~~~~~~~~~~~~~~~~~~~~~~~~~~~~~~~~~~~~~~
plot(result(:,1))
hold on
plot(result(:,2))
hold off
grid on
************************************************************

************************************************************
function omg=objrastr(x)
omg=20+5*x(1)^2+5*x(2)^2-10*(cos(2*pi*x(1))+cos(2*pi*x(2)));
************************************************************
```

In the GA implementation presented in the above listing, previously discussed operations are implemented in the following way.

The number of elite children is directly given (not as ratio), and those individuals are passed to the successive generation without any modification.

To control cross-over children two variables are used. Variable `cross` represents a ratio between cross-over children and overall population number.

This number needs to be even as it will form number of couples, representing parents, which will cross their genes (parameters) in order to form new children. In order to keep the number of individuals in the population constant during the evolution, from each couple two children are generated. Second variable used to control the crossing-over rule is `crossP` that represents the number of individuals (expressed in ratio of the overall population number) that will be considered as potential parents for crossing over. For example, if this ratio is defined as 0.4, it means that first 40% of the individuals will be considered as potential parents (i.e. the best 40% in terms of fitness function value). From these individuals a number of couples defined by the other variable is randomly selected. Obviously with this implementation repetition of the same individuals in different couples is possible.

The number of individuals that will mutate is already defined and is equal to what remains from the total number of individuals when previous two groups are subtracted. The rest of the mutation process is controlled by additional two variables: `mutR` and `mutrange`. The former one defines ratio of individuals that will be considered as those to whom a random mutation will be applied, while the latter represents a range of "jump" attributed to the parameters. For example if the `mutR` if set to 0.3 it means that from last 30% of the individuals (i.e. last in terms of fitness function value) a number of individuals will be selected according to previously calculated number of mutation children. This implies that also here the individuals may be repeated as they are selected randomly. On the other hand, `mutrange` refers to the maximum amount of mutation that can be attributed to each of the individuals. As previously illustrated this variable represents an important quantity as it directly influences the performance of the generic algorithm in the presence of local minima.

As for the stopping criteria in the present GA implementation only two of them are implemented, namely the total number of allowed generations is prescribed (variable `TOTGEN`) and the number of stalling generations is given (variable `TOTSTALL`), or number of consecutive generations in which any improvement in the objective function value is not achieved. With this stopping criteria the optimization will be terminated not later than `TOTGEN` generation, or even earlier if during the optimization for more than `TOTSTALL` generations there will be no improvement in the objective function.

During the optimization cycle matrix `result` serves to store the objective function value from the best individual, and the mean value of the objective function for each generation. These results are plotted at the end of the optimization in the graphical representation that is characteristic for the genetic algorithm optimizations.

Implemented GA can be used to solve optimization problem defined by Eq. 2.48 in order to verify the capability of the algorithm to solve difficult cases in terms of the number of local minima. This optimization problem is solved by prescribing the number of individuals in each simulation to be equal to 100, while the stopping criteria are defined by maximum allowed number of generations equal to 200, and stalling generation number equal to 80. By purpose, the initial population is generated in the zone of local minima (i.e. both parameters within the range

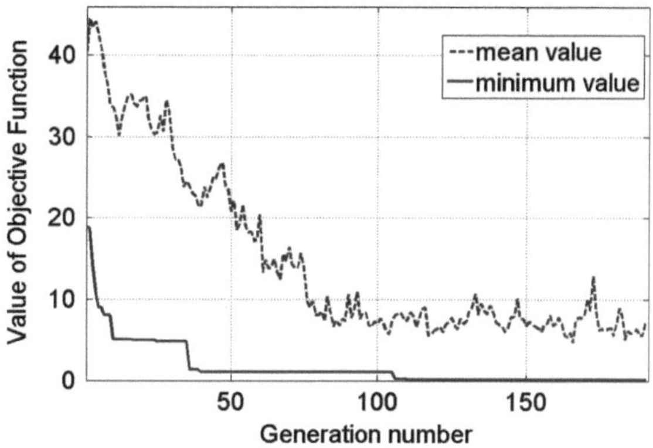

Fig. 2.28 Result of the GA optimization of Rastrigin's type function

[0.5, 2]). The amount of mutation prescribed (that also turned out to be sufficient given the performance of the algorithm) was equal to 2.5.

Figure 2.28 visualizes the result of this optimization. From the graph it may be observed how the value of objective function from the best individual in each generation descends monotonely, while keeping its value unchanged for some number of generations. On the other hand, the mean value of the objective function for each population fluctuates, as a result of random nature of the process which not necessarily improves the result of all the individuals. Nevertheless there is also a global descending nature of this graph which indicates the tendency of grouping the individuals in the zone of global minimum. The fact that mean value remains a bit detached from the minimum value shows a certain scattering of the individuals.

Even though our goal is not to bring *all* of the individuals to the global minimum value, since the result of optimization is anyhow represented by the best individual, still the following modification may contribute to overall improvement of the algorithm performance. Instead of having a constant value for the mutation amount, an alternative implementation can start with some larger value attributed to it (e.g. like 2.5 in this case), but at certain point changing it to somewhat smaller number. The reasoning behind this modification consists in the fact that this variable helps the algorithm to "jump" out of possible local minima, but after a certain number of generations it is reasonable to expect that most of the individuals would find themselves in the global minimum zone. Therefore, in order to have all the GA operations (i.e. elite selection, crossing-over and random mutation) working in the direction of improving the objective function within the present zone, a mutation range amount can be decreased.

By implementing a simple modification to previous code that will reduced mutation amount to one fourth of its initial value after the generation reaches number 50, the performance of the algorithm in fact are improved.

Result of the GA optimization with variable mutation amount is visualized in Fig. 2.29. For this particular case the improvement achieved by this modification is not large since the reduction in overall number of generations is not significant. The influence of this modification is evidenced also in the mean value of the objective

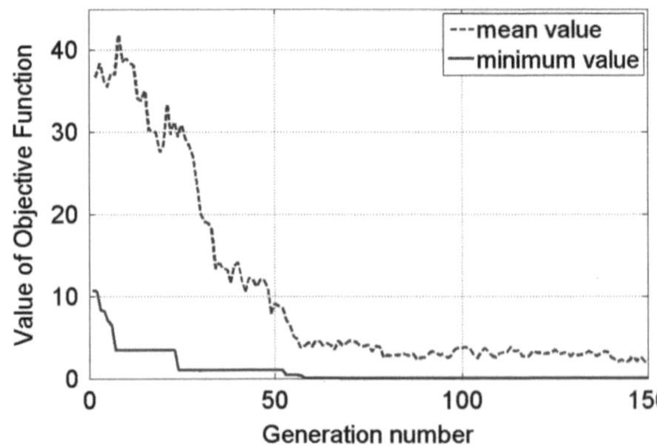

Fig. 2.29 Result of the GA optimization with variable mutation amount

function. At the end of the optimization, the mean value of the objective function for the last generation is smaller than in previous case, a circumstance that points out more dense distribution of the individuals in the zone of global minimum. This feature of the algorithm extends the probability of finding the individuals with smaller value of the objective function faster by the applied GA processes resulting therefore in potentially more effective optimization.

2.5 Summary

In this chapter some of the most frequently used optimization algorithms are discussed and their implementation into MATLAB codes is presented. These codes will be used further in this book as a part of inverse analysis procedures designed for the parameter identification.

First part of the chapter showed traditional, derivative based, optimization procedures. In particular two different strategies are discussed: Line search methods, and trust region methods. All of these algorithms have some strong and some weak points, and the text presented in this chapter attempted to point them out. For example, steepest descend line search algorithm has a good feature of being robust as with this approach a global convergence is guaranteed. However, as we could see, in some situations it can perform rather poor involving significantly larger number of function evaluations with respect to other algorithms. Newton direction line search is extremely powerful provided that Hessian matrix is positive definite, and that model function represents a good approximation of real objective function.

Trust region algorithms on the other hand use a different philosophy and at each step they minimize the model function subjected to a trust region constrain. This approach is particularly efficient if the objective function to be minimized is complicated, and therefore cannot be very accurately approximated by a quadratic form. The minimization is therefore restricted only to a nearby zone where it is trusted that the approximation is good enough.

Both of these groups of algorithms are solving minimization problem mathematically and therefore are sensitive to a presence of local minima (i.e. they cannot distinguish between local and global minimum). If the problem under consideration turns out to have a large number of local minima, all of these algorithms are ineffective. For these situations it is more convenient to use Genetic Algorithms.

A brief description on main principles together with a possible GA implementation is given in the second part of this chapter. In general, GA are involving more evaluations of the objective function. Within the problems of interest presented in this book, the evaluation of the objective function usually involves one simulation of the system response, so sometimes it may be a time consuming task. Therefore, it is computationally unjustifiable to use GA for problems that can be relatively successfully solved by traditional derivative based algorithms.

Optimization algorithm represents one part of the inverse analysis procedure. When it is required to design some parameter characterization procedure, an important issue is its robustness and stability. Examples treated in the chapter should serve to have an idea about the potentialities and limitations of particular optimization algorithms. The behavior of any algorithm depends strongly on the type of the function which should be optimized. Therefore, a general suggestion in selection of the particular algorithm that should be used within the procedure is to keep it as simple and robust as possible for the problem under consideration.

References

1. Giannessi, F.: Metodi matematici della programmazione. Problemi lineari e non lineare, Bologna (1982)
2. Nocedal, J., Wright, S.J.: Numerical Optimization. Springer, New York (2006)
3. Bonnans, J.F., Gilbert, J.C., Lemarechal, C., Sagastizabal, C.A.: Numerical Optimization – Theoretical and Practical Aspects. Springer, New York (2000)
4. Dussault, J.P.: Convergence of Implementable Descent Algorithms for Unconstrained Optimization. J Optimiz Theory App **104**(3), 739–745 (2000)
5. De Leone, R., Guadioso, M.: Stopping Criteria For Line Search Methods Without Derivatives. Math Program **30**(3), 285–300 (1984)
6. Powell, M.J.D.: A new algorithm for unconstrained optimization. Non-linear programming – Academic press: 34–65, New York (1970)
7. Branch, M.A., Coleman, T.F., Li, Y.: A Subspace, Interior, and Conjugate Gradient Method for Large-Scale Bound-Constrained Minimization Problems. SIAM J Sci Comput **21**(1), 1–23 (1999)
8. Byrd, R.H., Schnabel, R.B.: Approximate Solution of the Trust Region Problem by Minimization over Two-Dimensional Subspaces. Math Program **40**, 247–263 (1988)
9. Konar, A.: Artificial Intelligence and Soft Computing – Behavioral and Cognitive Modeling of Human Brain. CRC Press, New York (2000)
10. Poli, R., Langdon, W.B., McPhee, N.F.: A field guide to genetic programming. Published via http://lulu.com and freely available at http://www.gp-field-guide.org.uk, (With contributions by J. R. Koza (2008)

Chapter 3
Proper Orthogonal Decomposition and Radial Basis Functions for Fast Simulations

Proper Orthogonal Decomposition (POD) is a powerful method for low-order approximation of some high dimensional processes. It is widely used in the situations where model reduction is required. The most favorable feature of the method is its optimality: it provides the most efficient way of capturing the dominant components of high-dimensional processes with, sometimes surprisingly small number of "modes". This chapter will present an algorithm that combines POD with Radial Basis Functions (RBF) used for the interpolation of the data with previously reduced dimensionality by the POD.

3.1 Short History of Proper Orthogonal Decomposition

In the scientific community it is generally accepted that the Proper Orthogonal Decomposition was originally developed by Pearson [1] about 100 years ago as a tool for graphical analyses. However, there are actually some earlier examples of the development of this mathematical technique. Independently, Beltrami in 1873 and Jordan in 1874 derived the Singular Value Decomposition (SVD) which, as it will be shown later, reflects the same theory. This is a good example that, already in its early stages of the development, the technique showed its feature of attracting the attention of scientists from different fields, which will follow it throughout the century. This peculiar characteristic has as a consequence that the method seemed to be re-developed throughout the time by different authors under the different names. Hotelling in [2] developed the method in statistical data processing and probability theory, known as Hotelling transformation. Later, during 1940s, Karhunen [3] and Loeve [4] independently developed a theory regarding optimal series expansions of continuous-time stochastic processes, nowadays known as Karhunen-Loeve decomposition. Lumley [5] traced the idea of POD to the independent investigation not only of Karhunen and Loeve, but also Kosambi (1943), Pougachev (1953) and Obukhov (1954).

V. Buljak, *Inverse Analyses with Model Reduction*, Computational Fluid and Solid Mechanics, DOI 10.1007/978-3-642-22703-5_3,

Among many different names that this method bears, the most frequently used are the following:

- Principal Component Analyses (PCA)
- Karhunen-Loeve Decomposition (KLD)
- Singular Value Decomposition (SVD)

Recently, Liang et al. [6] showed the connections and equivalence of all the three methods, proving that all these names are practically refereeing to the same mathematical procedure.

Since its introduction the method has received much attention as a tool to analyze complex physical systems. During the years of development, POD has been used in variety of fields in science and engineering. Some of the most interesting examples are mentioned here

- Transient thermal analysis [7]
- Description of turbulent fluid flows [8]
- Structural dynamics [9]
- Signal processing and control theory [10]
- Damage detection [11]
- Human face recognition [12]
- Unsteady aerodynamics [13]

What makes this method so popular in applications where low-dimensional high-accuracy approximations are needed is its optimality. POD provides a basis for the modal decomposition of an ensemble of data, such as those obtained in the experiments or numerical simulations. Properties of POD suggest that it is a preferable basis in a sense that it is constructed to maximize the accuracy of the approximation. Therefore for the given set of data and for the given number of modes used to approximate the data there is no other basis which can give better approximation in a least square sense, as it will be shown later. In order to show how this goal is achieved, let us first see what is the main idea behind the approximation procedures.

3.2 Approximation

Let us suppose that we wish to approximate a function $U(x)$over some domain of interest denoted by Ω. The function can be written as a linear combination of some basis functions $\boldsymbol{\varphi}^i(x)$

$$U(x) \approx \sum_{i=1}^{M} a_i \cdot \boldsymbol{\varphi}^i(x) \tag{3.1}$$

with the reasonable expectation that the approximation becomes exact in the limit case as M approaches the infinity. The coefficients denoted by a_i are representing the unknown amplitudes of the expansion. Once the basis functions are chosen, the amplitudes' values are obtained by a minimization process, which, for the approximation in the least square sense is defined as

$$\left\| U(x) - \sum_{i=1}^{M} a_i \, \varphi^i(x) \right\|^{L^2} \rightarrow \min \tag{3.2}$$

where $\|\cdot\|^{L^2}$ denotes L^2-norm defined by

$$\|f(x)\|^{L^2} = \int_{\Omega} |f(x)|^2 d\Omega \tag{3.3}$$

Clearly, the representation of (3.1) is not unique for the same function. The choice of functions $\varphi^i(x)$ to form the "basis" for $U(x)$ is arbitrary, and to each selected basis, a different set of amplitudes is corresponding. In standard approximation approach it is left to the user to choose the basis functions. Usually from the experience one can say if for a certain function the basis should be constructed from the polynomial functions, trigonometric, exponential, or any other type of functions. In most cases the target accuracy of approximation (3.1) can be achieved when M is large enough, but there is no proof that the selected basis is the best one for the given function. Therefore, it is natural to seek the basis that, for the given number M (possible relatively small) will approximate the function $U(x)$ at the best possible way. The POD deals particularly with the choice of the functions $\varphi^i(x)$, and offers a tool to construct an optimal basis for the function in question.

3.2.1 POD Approximation

There is a considerable freedom in selecting the set of basis functions $\varphi^i(x)$, provided that it is complete and linearly independent. Taking the orthonormal set with the following property

$$\int_{\Omega} \varphi_{k1}(x) \cdot \varphi_{k2}(x) dx = \begin{cases} 1 & k_1 = k_2 \\ 0 & k_1 \neq k_k \end{cases} \tag{3.4}$$

gives some advantages. In this case, the determination of the amplitudes becomes relatively simple as then

$$a_i = \int_{\Omega} U(x) \cdot \varphi^i(x) d\Omega \tag{3.5}$$

This means that the amplitude a_i depends only on function $\varphi^i(x)$ and not on the others. Should the basis function be non-orthogonal, the determination of the amplitudes would require a solution of a set of linear equations.

Furthermore, this basis needs to be optimal, in a sense that for each value of M, the approximation should be as good as possible in terms of least square error (Eq. 3.2). In other words, the goal is to find a sequence of orthonormal functions such that the first two of these functions give the best possible two-term approximation, the first three the best possible three-term approximation and so on. Once found, these special ordered orthogonal functions are called the *proper orthogonal nodes* for the function $U(x)$ and the Eq. 3.1 is called the Proper Orthogonal Decomposition of $U(x)$.

3.3 Discrete POD Theory

Having in mind that the applications treated in this book are of numerical nature (or experimental one) only the discrete version of the POD theory will be presented. The discrete POD theory is usually found in the literature under the name of Principal Component Analyses (PCA), which would be the name used in this book, although some authors are also calling it discrete Karhunen-Loeve decomposition.

PCA is a method originally developed in statistics [1], which deals with random variables, and so the method itself is naturally of a discrete nature. The central idea of PCA is to reduce the dimensionality of a data set consisting of a large number of correlated variables, while retaining as much as possible of the variation present in the data set. This task is achieved by transforming the data to a new set of variables, called the Principal Components (PCs), which are uncorrelated, and which are ordered so that the first few retain most of the variation present in all of the original variables. Conversely, the last few PCs identify directions in which there is very little variation; that is, they identify near-constant linear relationship among the original variables. Once the PCs are identified, it is possible to reduce the dimensionality of the original space of variables, by keeping just the first few PCs, with still good accuracy of the approximation since PCA had ordered these new variables in such way that first components retain most of the variation of the original set.

If we consider relatively unrealistic, but simple case, where the original number of variables is equal to 2, it will allow us to have an easy to visualize geometrical interpretation of PCA. Treating the variables as components of vectors, each pair of variables will indicate one vector in 2D space. In such case, the term "correlated variables" would indicate that the vectors are parallel. Therefore, randomly distributed vectors, without any preferable orientation, like those visualized in Fig. 3.1a, represent visualization of uncorrelated variables.

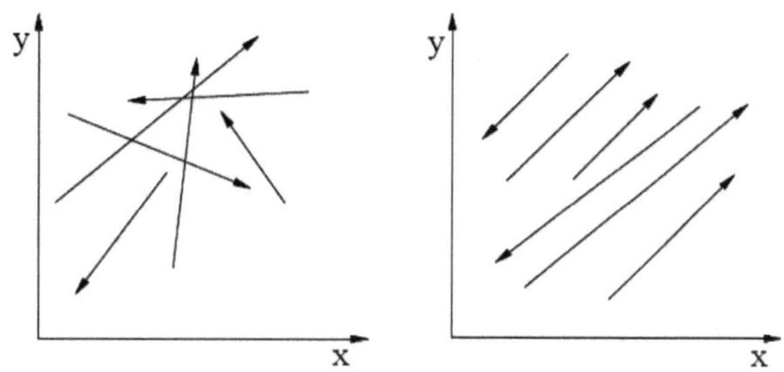

Fig. 3.1 Uncorrelated (**a**) and closely correlated (**b**) vectors

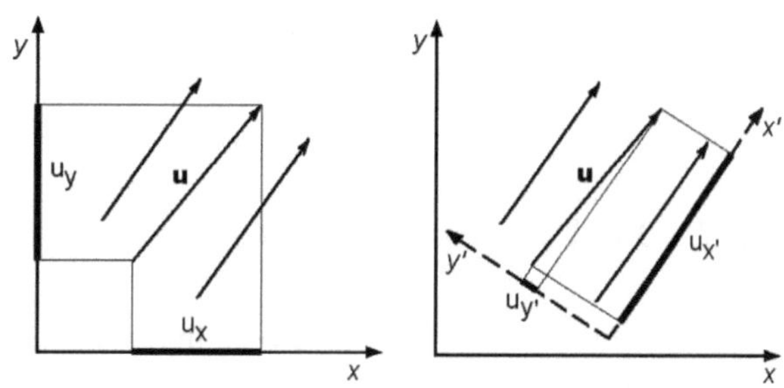

Fig. 3.2 One component approximation of closely correlated vectors in original and new coordinate basis

In this simple example, the reduction of dimensionality means that we would like to express our vectors just with one component (to cut-down the dimensionality from two to one). Let us imagine that we have a set of closely correlated vectors (almost parallel) like those visualized in Fig. 3.2. Considering the orientation of the vectors, it is quit obvious that this, one component approximation in original coordinate system will introduce a significant error, since both components of all the vectors have comparable magnitudes. On the other hand it is possible to introduce a new coordinate system with axes rotated by approximately 45^0 (Fig. 3.2 right-hand side). Obviously, this new coordinate system has been chosen in such way, that one coordinate is dominant, and therefore in this system, by neglecting the other component the loss of data will be much less. In the limit case when all the vectors are fully correlated (parallel), by choosing axis *x'* to be parallel to the vectors, the lost data would be zero, since all the vectors in the new coordinate system are parallel to one axis, having the projection to the other one equal to zero. In all other cases, the approximation would introduce some error, since at least some of the vectors will have a projection to the other axis.

PCA defines the new coordinate system in order to minimize this loss of the data in average, least square sense for all the vectors. Therefore, if the axis *x'* is chosen in such way that for the given vector set it will give the best possible one component approximation, then this new basis represents a Proper Orthogonal Basis for the vectors in set.

The Principal Component Analysis therefore in 2D space can be interpreted as a rotation of the coordinate system to be as parallel as possible to the set of vectors. This concept can be extended to the vectors of arbitrary dimensionality.

As analogy to the continuous POD theory, where the goal is to find special ordered orthogonal functions (called the *proper orthogonal nodes*), here, the goal is to find orthogonal *directions* in N-dimensional space (where N is the original dimensionality of the vectors in set). Also here these directions need to be specially ordered in a way that, the first one will give the best possible one component approximation of vectors, the first two the best possible two component approximation and so on. Once found, they provide a favorable basis for reduction of the dimensionality of vectors in set, due to their optimal property, guaranteeing that there cannot be any other basis that can have better approximation for any selected number of reduced components.

There are different ways to construct POD basis for the vectors in set, and in what follows three of them will be presented in details.

3.3.1 PCA Derivation by Minimizing the Error of Approximation

The derivation of PCA presented both in this and subsequent sections will refer to an arbitrary case of vectors with dimensionality $N > 2$, as this is the case of practical interest both in the applications presented in this book and in general applications of reduction of dimensionality.

The derivation of PCA can proceed starting directly from the definition, namely from minimizing the error of approximation. Let us assume that we have a set of N-dimensional vectors (say, M of them), collected in matrix $\mathbf{U}$. Our goal is to find the most accurate representation in some subspace W with the dimension of $K < N$. If we denote by $\boldsymbol{\varphi}_1, \boldsymbol{\varphi}_{2,...,}\boldsymbol{\varphi}_K$ the orthonormal basis of W, then each vector from the original set can be written as

$$\mathbf{u}_i \approx \sum_{j=1}^{K} \bar{a}_{ij} \cdot \boldsymbol{\varphi}_j, i = 1, ...N \tag{3.6}$$

where $\bar{a}_{ij}$ are amplitudes corresponding to i^{th} vector in new subspace W, and $\bar{\Phi}$ is matrix that collects all the orthonormal basis of the subspace $\boldsymbol{\varphi}_j$. Note that here both amplitudes and matrix of basis are marked with the bar, which will be used throughout the book for the symbols corresponding to the subspace of the reduced dimensionality. This can be written in the matrix form as

$$\mathbf{u}_i \approx \bar{\Phi} \cdot \bar{\mathbf{a}}_i, i = 1, ...N \tag{3.7}$$

where symbol $\approx$ suggest that it represents an approximation of the vector $\mathbf{u}_i$. The error of this approximation in least square sense can be written as

$$error = \left\| \mathbf{u}_i - \sum_{j=1}^{K} \bar{a}_{ij} \cdot \boldsymbol{\varphi}_j \right\|^{L^2} \tag{3.8}$$

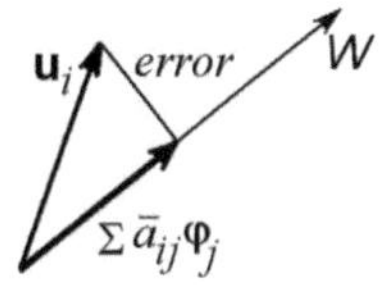

Fig. 3.3 Geometrical interpretation of the error of approximation

Geometrically, this error can be interpreted as a squared perpendicular distance of $\mathbf{u}_i$ from the subspace W (Fig. 3.3).

Equation 3.8 represents the error just for i^{th} vector. To have the overall error for all the vectors we need to make a summation over all of them. Denoting with E the total error we can write

$$E = \sum_{i=1}^{N} \left\| \mathbf{u}_i - \sum_{j=1}^{K} \bar{a}_{ij} \cdot \boldsymbol{\varphi}_j \right\|^{L^2} \tag{3.9}$$

Therefore, the total error represents summation of the squared perpendicular distances of all the vectors from the subspace W. Now the goal is to find the set of orthonormal basis $\boldsymbol{\varphi}_1, \boldsymbol{\varphi}_{2,\ldots,}\boldsymbol{\varphi}_K$ that minimizes this error. In order to minimize E we need to take the first derivatives with reference to the all unknowns that are uniquely defining the approximation, namely with respect to both orthonormal basis and amplitudes. Before doing that, let us first write Eq. 3.9 in a slightly different way.

$$\begin{aligned} E &= \sum_{i=1}^{N} \|\mathbf{u}_i\|^2 - 2\sum_{i=1}^{N} \mathbf{u}_i^T \cdot \left(\sum_{j=1}^{K} a_{ij} \cdot \boldsymbol{\varphi}_j \right) + \sum_{i=1}^{N}\sum_{j=1}^{K} a_{ij}^2 = \\ &= \sum_{i=1}^{N} \|\mathbf{u}_i\|^2 - 2\sum_{i=1}^{N}\sum_{j=1}^{K} a_{ij}\mathbf{u}_i^T\boldsymbol{\varphi}_j + \sum_{i=1}^{N}\sum_{j=1}^{K} a_{ij}^2 \end{aligned} \tag{3.10}$$

Now we can find partial derivatives, first with respect to a_{lm}

$$\frac{\partial E}{\partial a_{lm}} = -2\mathbf{u}_l^T\boldsymbol{\varphi}_m + 2a_{lm} \tag{3.11}$$

From (3.11) follows that the optimal value for a_{lm} is equal to

$$a_{lm} = \mathbf{u}_l^T\boldsymbol{\varphi}_m \tag{3.12}$$

Substituting the optimal value back to (3.10) it becomes

$$E = \sum_{i=1}^{N} \|\mathbf{u}_i\|^2 - 2\sum_{i=1}^{N}\sum_{j=1}^{K} (\mathbf{u}_i^T\boldsymbol{\varphi}_j)\mathbf{u}_i^T\boldsymbol{\varphi}_j + \sum_{i=1}^{N}\sum_{j=1}^{K} (\mathbf{u}_i^T\boldsymbol{\varphi}_j)^2 \tag{3.13}$$

Equation 3.13 can be written in a different way. Since the value of scalar that results from vector product of two vectors can be written in two ways, namely

$$\mathbf{u}_i^T \boldsymbol{\varphi}_j = \boldsymbol{\varphi}_j^T \mathbf{u}_i \tag{3.14}$$

mid-term of (3.13) can be rewritten as

$$2\sum_{i=1}^{N}\sum_{j=1}^{K}(\mathbf{u}_i^T \boldsymbol{\varphi}_j)\mathbf{u}_i^T \boldsymbol{\varphi}_j = 2\sum_{i=1}^{N}\sum_{j=1}^{K}\boldsymbol{\varphi}_j^T \mathbf{u}_i \mathbf{u}_i^T \boldsymbol{\varphi}_j \tag{3.15}$$

On the other hand, the last term of (3.13) can be expressed as

$$\sum_{i=1}^{N}\sum_{j=1}^{K}(\mathbf{u}_i^T \boldsymbol{\varphi}_j)^2 = \sum_{i=1}^{N}\sum_{j=1}^{K}(\mathbf{u}_i^T \boldsymbol{\varphi}_j)^T(\mathbf{u}_i^T \boldsymbol{\varphi}_j) = \sum_{i=1}^{N}\sum_{j=1}^{K}\boldsymbol{\varphi}_j^T \mathbf{u}_i \mathbf{u}_i^T \boldsymbol{\varphi}_j \tag{3.16}$$

Substituting (3.15) and (3.16) back to (3.13) the total error to be minimized obtains the following form

$$E = \sum_{i=1}^{N}\|\mathbf{u}_i\|^2 - \sum_{i=1}^{N}\sum_{j=1}^{K}(\mathbf{u}_i^T \boldsymbol{\varphi}_j)^2 \tag{3.17}$$

or, once again rewritten in the form like (3.15) it becomes

$$E = \sum_{i=1}^{N}\|\mathbf{u}_i\|^2 - \sum_{j=1}^{K}\boldsymbol{\varphi}_j^T\left(\sum_{i=1}^{N}\mathbf{u}_i\mathbf{u}_i^T\right)\boldsymbol{\varphi}_j = \sum_{i=1}^{N}\|\mathbf{u}_i\|^2 - \sum_{j=1}^{K}\boldsymbol{\varphi}_j^T\mathbf{C}\boldsymbol{\varphi}_j \tag{3.18}$$

where $\mathbf{C}$ is, so-called covariance matrix defined as

$$\mathbf{C} = \mathbf{U}\mathbf{U}^T \tag{3.19}$$

By observing the equation of total error to be minimized expressed by (3.18) it is obvious that it has the following form

$$E = \beta - \sum_{j=1}^{K}\boldsymbol{\varphi}_j^T\mathbf{C}\boldsymbol{\varphi}_j \tag{3.20}$$

where scalar β is a constant depending on original set of vectors (not on the choice of the new basis). Recalling that the goal here was to find a new basis with reduced space that minimizes the error of the approximation for the given dimensionality K, obviously scalar β is uninfluenced by the selection of this new basis, therefore it follows that in order to minimize the error, the second term in (3.20) needs to be maximized

$$\sum_{j=1}^{K} \boldsymbol{\varphi}_j^T \mathbf{C} \boldsymbol{\varphi}_j \rightarrow \max \tag{3.21}$$

under the constraint of orthonormaility of the new basis

$$\boldsymbol{\varphi}_j^T \boldsymbol{\varphi}_j = 1, j = 1, \ldots K \tag{3.22}$$

In order to solve this constrained maximization problem, one can use Lagrange multipliers method which converts the problem into

$$\sum_{j=1}^{K} \boldsymbol{\varphi}_j^T \mathbf{C} \boldsymbol{\varphi}_j - \sum_{j=1}^{K} \lambda_j \left(\boldsymbol{\varphi}_j{}^T \boldsymbol{\varphi}_j - 1 \right) \rightarrow \max \tag{3.23}$$

Now let us compute the derivatives of (3.23) with respect to $\boldsymbol{\varphi}_j$. If $\mathbf{x}$ is a vector, f is some function of $\mathbf{x}$, denoting the first derivative with respect to $\mathbf{x}$ by

$$\frac{d}{d\mathbf{x}} f(\mathbf{x}) = \begin{bmatrix} \frac{\partial f}{\partial x_1} \\ \cdots \\ \frac{\partial f}{\partial x_d} \end{bmatrix} \tag{3.24}$$

it can be shown that

$$\frac{d}{d\mathbf{x}} \left(\mathbf{x}^T \mathbf{x} \right) = 2\mathbf{x} \tag{3.25}$$

and that for any symmetric matrix $\mathbf{A}$ it holds

$$\frac{d}{d\mathbf{x}} \left(\mathbf{x}^T \mathbf{A} \mathbf{x} \right) = 2\mathbf{A}\mathbf{x} \tag{3.26}$$

Having this in mind, since matrix $\mathbf{C}$ is symmetric, by equalizing the first derivatives of (3.23) with respect to $\boldsymbol{\varphi}_j$ with zero we have

$$\frac{d}{d\boldsymbol{\varphi}_j} \left(\sum_{j=1}^{K} \boldsymbol{\varphi}_j \mathbf{C} \boldsymbol{\varphi}_j - \sum_{j=1}^{K} \lambda_j \left(\boldsymbol{\varphi}_j{}^T \boldsymbol{\varphi}_j - 1 \right) \right) = 2\mathbf{C}\boldsymbol{\varphi}_j - 2\lambda_j \boldsymbol{\varphi}_j = 0 \tag{3.27}$$

The last equation is satisfied if $\boldsymbol{\varphi}_j$ is eigenvector and λ_j is corresponding eigenvalue of matrix $\mathbf{C}$ since it requires that

$$\mathbf{C}\boldsymbol{\varphi}_j = \lambda_j \boldsymbol{\varphi}_j \tag{3.28}$$

Substituting (3.28) into (3.18) overall error of the approximation becomes

$$E = \sum_{i=1}^{N} \|\mathbf{u}_i\|^2 - \sum_{j=1}^{K} \boldsymbol{\varphi}_j^T \mathbf{C} \boldsymbol{\varphi}_j = \sum_{i=1}^{N} \|\mathbf{u}_i\|^2 - \sum_{j=1}^{K} \boldsymbol{\varphi}_j^T \lambda_j \boldsymbol{\varphi}_j$$
$$= \sum_{i=1}^{N} \|\mathbf{u}_i\|^2 - \sum_{j=1}^{K} \lambda_j \tag{3.29}$$

Recalling that the first term is a constant it results that the error of approximation is minimized if the new basis is constructed of K eigenvectors that are corresponding to the first K largest eigenvalues of covariance matrix $\mathbf{C}$

$$\bar{\Phi} = [\boldsymbol{\varphi}_j],\ j = 1, \ldots K \tag{3.30}$$

In a trivial case when all the eigenvectors of matrix $\mathbf{C}$ are taken to construct the subspace W, there will be no error of the approximation. In fact, there will be no approximation at all, since in that case all the vectors $\mathbf{u}_i$ are just expressed in a different coordinate basis. Approximation in any other subspace, that uses smaller number of directions will introduce an error, and from (3.29) it is obvious that the introduce error is directly proportional to the summation of the eigenvalues corresponding to the neglected directions $\sum_{i=K+1}^{N} \lambda_j$. Usually the error of approximation is expressed using the ratio between kept eigenvalues and all of them, namely

$$\frac{\sum_{i=1}^{K} \lambda_i}{\sum_{i=1}^{M} \lambda_i} \tag{3.31}$$

Example 3.1. Consider a set of four three dimensional vectors that are collected in a matrix

$$\mathbf{U} = \begin{bmatrix} 1 & 5 & 3 & 3 \\ 1 & 4 & 4 & 3 \\ 1 & 5 & 5 & 4 \end{bmatrix}$$

Construct POD basis and compare the accuracy between one-component and two-component approximations with the original set of vectors

---

Solution:

First we need to compute covariance matrix $\mathbf{C}$ *that is equal to*

$$\mathbf{C} = \begin{bmatrix} 1 & 5 & 3 & 3 \\ 1 & 4 & 4 & 3 \\ 1 & 5 & 5 & 4 \end{bmatrix} \cdot \begin{bmatrix} 1 & 1 & 1 \\ 5 & 4 & 5 \\ 3 & 4 & 5 \\ 3 & 3 & 4 \end{bmatrix} = \begin{bmatrix} 44 & 42 & 53 \\ 42 & 42 & 53 \\ 53 & 53 & 67 \end{bmatrix}$$

(continued)

The solution of eigenvalue problem gives us eigenvectors and corresponding eigenvalues of matrix $\mathbf{C}$ *equal to*

$$\Phi = \begin{bmatrix} 0.5325 & 0.8462 & -0.0183 \\ 0.5255 & -0.3135 & -0.3135 \\ 0.6636 & -0.4308 & -0.6116 \end{bmatrix}, \ \lambda_1 = 151.49; \ \lambda_2 = 1.46; \ \lambda_3 = 0.045$$

Note that there is fast trend in dropping of the magnitudes of eigenvalues that suggest strong correlation between the vectors (since vectors in this set are close to parallel)

Now let us compute the amplitudes in this new basis. Having in mind orthonormality of the new basis, the matrix of amplitudes $\mathbf{A}$ *is computed by*

$$\mathbf{A} = \Phi^T \mathbf{U} = \begin{bmatrix} 0.5325 & 0.5255 & 0.6636 \\ 0.8462 & -0.3135 & -0.4308 \\ -0.0183 & 0.7909 & -0.6116 \end{bmatrix} \begin{bmatrix} 1 & 5 & 3 & 3 \\ 1 & 4 & 4 & 3 \\ 1 & 5 & 5 & 4 \end{bmatrix} =$$

$$= \begin{bmatrix} 1.7215 & 8.0822 & 7.0172 & 5.8282 \\ 0.1019 & 0.8232 & -0.8693 & -0.1250 \\ 0.1610 & 0.0142 & 0.0508 & -0.1285 \end{bmatrix}$$

As it was suggested by fast drop of eigenvalues, vectors are correlated, and the new coordinate basis is oriented such that its first direction is similar to the direction of the vectors. This can be seen from the amplitudes, since all of the vectors have the first amplitude one order of magnitude larger than the second one. It can be also seen that some of the vectors from original matrix $\mathbf{U}$*, for instance the first one, have the third amplitude larger than the second one. It means that this vector alone would actually have better two-component approximation if the first and the third direction should be used. Nevertheless, criterion for error minimization is derived to be satisfied in average least square sense for all the vectors in set, and of course it doesn't have to be satisfied for all vectors separately.*

One component approximation is constructed by keeping just the first direction and corresponding amplitudes, namely

$$\mathbf{U} \approx \bar{\mathbf{\Phi}}_1 \bar{\mathbf{A}}_1 = \begin{bmatrix} 0.5325 \\ 0.5255 \\ 0.6636 \end{bmatrix} [1.7215 \quad 8.0822 \quad 7.0172 \quad 5.8282]$$

$$= \begin{bmatrix} 0.9167 & 4.3037 & 3.7366 & 3.1034 \\ 0.9046 & 4.2468 & 3.6872 & 3.0625 \\ 1.1424 & 5.3633 & 4.6566 & 3.8676 \end{bmatrix}$$

while the two-component approximation is equal to

(continued)

$$\mathbf{U} \approx \bar{\mathbf{\Phi}}_2 \bar{\mathbf{A}}_2 = \begin{bmatrix} 0.5325 & 0.8462 \\ 0.5255 & -0.3135 \\ 0.6636 & -0.4308 \end{bmatrix} \begin{bmatrix} 1.7215 & 8.0822 & 7.0172 & 5.8282 \\ 0.1019 & 0.8232 & -0.8693 & -0.1250 \end{bmatrix}$$

$$= \begin{bmatrix} 1.0029 & 5.0003 & 3.0009 & 2.9976 \\ 0.8726 & 3.9887 & 3.9598 & 3.1016 \\ 1.0985 & 5.0087 & 5.0311 & 3.9214 \end{bmatrix}$$

Comparing the approximations to the starting matrix **U** *a small difference can be noticed.*

3.3.2 PCA Derivation Based on Correlation Matrix

PCA is oriented to reduce dimensionality of the correlated data. In fact the more correlated the vectors are, the larger reduction of the space can be obtained, as we already have seen. Therefore, an alternative way to derive the PCs could be by using correlation, rather than covariance matrix. In [14] Jolliffe showed a set of advantages and disadvantages of the derivation based on the correlation matrix over the one that uses covariance matrix. However, most of the discussions given there are considering statistical framework, which is not the scope of this book. Here we will limit ourselves just on showing that this alternative approach leads to exactly the same POD basis, for sometimes smaller computational cost.

Let us recall that the goal is to reduce the dimensionality of a data set by deriving a small number of linear combinations (called principal components) of a set of variables that retain as much of the variation in the original variables as possible. Assuming that the matrix **U** collects original set of M, N-dimensional vectors $\mathbf{u}_i$, in terms of the requirements of PCA let the first principal component be a linear combination of each of them

$$\boldsymbol{\varphi}_1 = \mathbf{U} \cdot \mathbf{v}_1 \tag{3.32}$$

where the unknown coefficients of this combination are forming an M-dimensional vector $\mathbf{v}_1$.

By assuming that the vector $\mathbf{v}_1$ is a unit vector, namely

$$[\mathbf{v}_1]^T \cdot \mathbf{v}_1 = 1 \tag{3.33}$$

the direction along which there is the largest variance of all the vectors collected in matrix **U** will maximize the length of vector $\boldsymbol{\varphi}_1$

$$[\boldsymbol{\varphi}_1]^T \cdot \boldsymbol{\varphi}_1 \rightarrow \max \tag{3.34}$$

This criterion is quit obvious, since the unknown direction is defined as a linear combination of all the vectors collected in matrix **U**. Therefore, for fixed matrix **U**, the unknown direction is uniquely defined with the set of coefficients collected in vector $\mathbf{v}_{1.}$ Changing these coefficients under the constraint that they should form a unit vector will obviously change not only the direction of $\boldsymbol{\varphi}_1$, but also its length. The direction that matches the one with the largest variance of all the vectors from the set will have the biggest length of vector $\boldsymbol{\varphi}_1$. Substituting (3.32) into (3.34) we have

$$[\mathbf{v}_1]^T \cdot \mathbf{U}^T \cdot \mathbf{U} \cdot \mathbf{v}_1 \rightarrow \max \tag{3.35}$$

which becomes maximization problem with the constraint given by (3.33). To solve this constrained problem, once again the Lagrange multipliers method can be used. With this, the new formulation of the problem is defined by

$$[\mathbf{v}_1]^T \cdot \mathbf{D} \cdot \mathbf{v}_1 + \lambda_1 \left(1 - [\mathbf{v}_1]^T \cdot \mathbf{v}_1\right) \rightarrow \max \tag{3.36}$$

where

$$\mathbf{D} = \mathbf{U}^T \cdot \mathbf{U} \tag{3.37}$$

is so-called modified correlation matrix. It is called modified, since in general vectors collected in matrix **U** don't have to be unit vectors. In the case when all the columns of matrix **U** are normalized to be a unit vectors, matrix **D** will be exactly the correlation matrix of vectors $\mathbf{u}_i$ (namely its element d_{ij} will be a correlation coefficient between column i and column j of matrix **U**). The correlation coefficient is within the range –1 to 1. Value 1 means that the two vectors are fully correlated (linearly dependant). Value –1 means that the two vectors are fully correlated and opposite by the sign. Value 0 means that the vectors are uncorrelated (they are orthogonal). All the values in between are giving quantitative information about the correlation between the two vectors. The diagonal elements of the correlation matrix are of course 1, since they give correlation coefficients for vectors $\mathbf{u}_i$ with themselves. The correlation matrix is symmetric since the correlation of column i with column j is the same as the correlation of column j with column i. In more general case, when vectors $\mathbf{u}_i$ are not unit vectors, modified correlation matrix will not have elements within the range –1 to 1, but it will be still symmetric and positive-semi-definite.

The maximization problem given by (3.36) needs to be solved for unknown coefficients collected in vector $\mathbf{v}_1$, that are uniquely defining the unknown direction. Differentiation of (3.36) with respect to the unknown coefficients collected in vector $\mathbf{v}_1$ leads to

$$\mathbf{D} \cdot \mathbf{v}_1 - \lambda_1 \cdot \mathbf{v}_1 = 0 \tag{3.38}$$

This shows that the solutions of the constrained problem (3.35) and (3.33) are the eigenvalue and the corresponding eigenvector of the matrix **D**. It is known from the linear algebra that the eigenvectors form a set of orthonormal vectors

$$[\mathbf{v}_i]^T \cdot \mathbf{v}_j = \delta_{ij} \tag{3.39}$$

where δ_{ij} is a Kronecker delta. Having in mind this orthonormality, multiplying Eq. 3.38 by $[\mathbf{v}_1]^T$gives

$$\left[\mathbf{v}^1\right]^T \cdot \mathbf{D} \cdot \mathbf{v}^1 = \lambda_1 \tag{3.40}$$

Comparing the last equation with (3.35) it is obvious that the solution of the maximization problem is obtained by taking the largest eigenvalue and the corresponding eigenvector.

The next direction can be found in an analogous manner, which will lead to the following equation

$$\mathbf{D} \cdot \mathbf{v}_2 - \lambda_2 \cdot \mathbf{v}_2 = 0 \tag{3.41}$$

and the second direction should be computed as

$$\boldsymbol{\varphi}_2 = \mathbf{U} \cdot \mathbf{v}_2 \tag{3.42}$$

Finally, the general expression to construct the ith optimal direction can be written as

$$\mathbf{D} \cdot \mathbf{v}_i - \lambda_2 \cdot \mathbf{v}_i = 0 \tag{3.43}$$

and

$$\boldsymbol{\varphi}_i = \mathbf{U} \cdot \mathbf{v}_i \tag{3.44}$$

with the eigenvalues taken in decreasing order.

This derivation however is not leading to normalized basis $\boldsymbol{\varphi}_i$. Up to know, only vector $\mathbf{v}_i$ was constrain to be a unit vector, and so the maximization procedure in general case will result with basis vectors $\boldsymbol{\varphi}_i$ of arbitrary length. Since it is convenient to work with orthonormal basis, the entries of vectors $\mathbf{v}_i$ should be scaled

$$\widehat{\mathbf{v}}_i = \mathbf{v}_i \cdot z_i \tag{3.45}$$

where z_i is an unknown normalizing factor.

The general formula for i^{th} optimal orthonormal direction is now given in the following form

$$\widehat{\boldsymbol{\varphi}}_i = \mathbf{U} \cdot \widehat{\mathbf{v}}_i = \mathbf{U} \cdot \mathbf{v}_i \cdot z_i \tag{3.46}$$

where terms with the hat are corresponding to normalized unit length directions. Therefore, the unknown normalizing parameter is obtained by forcing the fulfillment of the criterion

$$\left[\widehat{\boldsymbol{\varphi}}_i\right]^T \widehat{\boldsymbol{\varphi}}_i = 1 \tag{3.47}$$

Substituting (3.46) into (3.47) the criterion becomes

$$[\mathbf{U} \cdot \mathbf{v}_i \cdot z_i]^T \mathbf{U} \cdot \mathbf{v}_i \cdot z_i = 1 \tag{3.48}$$

After regrouping we will have

$$\mathbf{v}_i^T z_i \mathbf{U}^T \mathbf{U} \mathbf{v}_i z_i = z_i^2 \mathbf{v}_i^T \mathbf{D} \mathbf{v}_i = 1 \tag{3.49}$$

By using (3.43), (3.49) can be further written as

$$z_i^2 \mathbf{v}_i^T \lambda_i \mathbf{v}_i = z_i^2 \lambda_i \mathbf{v}_i^T \mathbf{v}_i = 1 \tag{3.50}$$

Since vector $\mathbf{v}_i$ is a unit vector the last equation finally becomes

$$z_i^2 \lambda_i = 1 \tag{3.51}$$

which is now giving the value of normalization factor equal to

$$z_i = \lambda_i^{-1/2} \tag{3.52}$$

It was already mentioned that matrix $\mathbf{D}$ $[M \times M]$ is symmetric positive semi-definite so all M eigenvalues are non-negative, real numbers.

Finally, we can write that the orthonormal POD basis has columns defined by

$$\widehat{\boldsymbol{\varphi}}_i = \mathbf{U} \cdot \mathbf{v}_i \cdot \lambda_i^{-\frac{1}{2}} \tag{3.53}$$

and sorted in descending order of eigenvalues λ_i in a matrix

$$\Phi = \left[\widehat{\boldsymbol{\varphi}}_i\right], i = 1\ ,\dots\ M \tag{3.54}$$

It should be emphasized that the number of columns is not necessary equal to M. The number of columns will be equal to the number on non-zero eigenvalues of matrix $\mathbf{D}$. Matrix $\mathbf{D}$ is of the size $M \times M$, but it's rank can be lower than M. In the case when $N < M$ the rank of matrix $\mathbf{D}$ will be N (or even lower) and therefore, the new basis will be of $N \times N$ size. In the case of linearly dependant columns the rank will be even lower than N. This feature will be illustrated in the following example.

Example 3.2. Consider the same set of vectors as the one given in Example 3.1. Show that derivation of the POD basis starting from modified correlation matrix **D** leads to the same results as the one that uses covariance matrix.

Solution:
Let us first compute matrix **D** *that is equal to*

$$\mathbf{C} = \begin{bmatrix} 1 & 1 & 1 \\ 5 & 4 & 5 \\ 3 & 4 & 5 \\ 3 & 3 & 4 \end{bmatrix} \cdot \begin{bmatrix} 1 & 5 & 3 & 3 \\ 1 & 4 & 4 & 3 \\ 1 & 5 & 5 & 4 \end{bmatrix} = \begin{bmatrix} 3 & 14 & 12 & 10 \\ 14 & 66 & 56 & 47 \\ 12 & 56 & 50 & 41 \\ 10 & 47 & 41 & 34 \end{bmatrix}$$

The solution of eigenvalue problem for matrix **D** *gives the eigenvectors (collected as columns of matrix* **V***) and corresponding eigenvalues (given in a diagonal matrix* **L***)*

$$\mathbf{V} = \begin{bmatrix} 0.1399 & -0.0844 & -0.7572 & 0.6325 \\ 0.6566 & -0.6814 & -0.0669 & -0.3162 \\ 0.5701 & 0.7196 & -0.2390 & -0.3162 \\ 0.4735 & 0.1035 & 0.6042 & 0.6325 \end{bmatrix}$$

$$\mathbf{L} = \begin{bmatrix} 151.49 & 0 & 0 & 0 \\ 0 & 1.46 & 0 & 0 \\ 0 & 0 & 0.045 & 0 \\ 0 & 0 & 0 & 0 \end{bmatrix}$$

It may be observed that the eigenvalues are the same as in the example 3, and that, even though matrix **D** *is 4×4 matrix, it's rank is 3 since the fourth eigenvalue is equal to zero. In the present context this is expectable since the original size of the vector space was 3 and therefore also the new space can have maximum size equal to 3. Consequently, the fourth eigenvalue needs to be equal to zero.*

Considering just the first three eigenvalues and corresponding eigenvectors, the POD basis is constructed using (3.53) which written in a matrix form is equal to

$$\Phi = \mathbf{UVZ} = \begin{bmatrix} 1 & 5 & 3 & 3 \\ 1 & 4 & 4 & 3 \\ 1 & 5 & 5 & 4 \end{bmatrix} \begin{bmatrix} 0.1399 & -0.0844 & -0.7572 \\ 0.6566 & -0.6814 & -0.0669 \\ 0.5701 & 0.7196 & -0.2390 \\ 0.4735 & 0.1035 & 0.6042 \end{bmatrix}$$

$$\times \begin{bmatrix} 0.081 & 0 & 0 \\ 0 & 0.828 & 0 \\ 0 & 0 & 4.702 \end{bmatrix}$$

$$= \begin{bmatrix} 0.5325 & -0.8462 & 0.0183 \\ 0.5255 & 0.3135 & -0.7909 \\ 0.6636 & 0.4308 & 0.6116 \end{bmatrix}$$

(continued)

where matrix **Z** *is the diagonal matrix with normalization factors given by (3.52).*

Having in mind orthonormality of the new basis, the matrix of amplitudes **A** *is computed by*

$$\mathbf{A} = \Phi^T\mathbf{U} = \begin{bmatrix} 0.5325 & 0.5255 & 0.6636 \\ -0.8462 & 0.3135 & 0.4308 \\ 0.0183 & -0.7909 & 0.6116 \end{bmatrix} \begin{bmatrix} 1 & 5 & 3 & 3 \\ 1 & 4 & 4 & 3 \\ 1 & 5 & 5 & 4 \end{bmatrix}$$
$$= \begin{bmatrix} 1.7215 & 8.0822 & 7.0172 & 5.8282 \\ -0.1019 & -0.8232 & 0.8693 & 0.1250 \\ -0.1610 & -0.0142 & -0.0508 & 0.1285 \end{bmatrix}$$

As already anticipated, the two methods are resulting in the same POD basis and therefore, in the same approximations. Which one of the two will be used is left on the user to choose. However, it can be noted that matrix **C** is of $N \times N$ size, while matrix **D** is $M \times M$. Since it is always better to work on smaller matrix, it is therefore reasonable to choose approach based on matrix **C** in the cases when $N < M$ and the approach based on matrix **D** in the opposite case.

3.3.3 Construction of POD Basis: Singular Value Decomposition Approach

The third method of derivation of the POD basis that will be presented here is the Singular Value Decomposition (SVD). SVD can be viewed as the extension of the eigenvalue decomposition for the case of non-square matrices. For more detailed discussion on SVD, reader may refer to [15]. In general, this decomposition states, that for any real rectangular $N \times M$ matrix **U**, there exists orthogonal matrices, $N \times N$ $\mathbf{V}^1$ and $M \times M$ $\mathbf{V}^2$ such that

$$\left(\mathbf{V}^1\right)^T \mathbf{U}\mathbf{V}^2 = \mathbf{S} \tag{3.55}$$

where **S** is rectangular $N \times M$ matrix with all the elements equal to zero except those on the main diagonal, which are equal to

$$s_{ii} = \sigma_i, \sigma_1 > \sigma_2 > \ldots > \sigma_r > 0, r = \min(M,N) \tag{3.56}$$

σ_i are called singular values of matrix **U**, while columns of matrix $\mathbf{V}^1$ are called left singular vectors, and columns of matrix $\mathbf{V}^2$ right singular vectors. A non-negative number σ is called a singular value of real rectangular matrix **U** if and only if there exist a unit-length N-dimensional vector $\mathbf{v}^1$ and a unit-length M-dimensional vector $\mathbf{v}^2$ such that

$$\mathbf{U}\mathbf{v}^2 = \sigma\mathbf{v}^1 \text{ and } \mathbf{U}^T\mathbf{v}^1 = \sigma\mathbf{v}^2 \tag{3.57}$$

The rank of matrix $\mathbf{U}$ equals the number of non-zero singular values. However, in the presence of numerical or experimental noises, the determination of matrix rank becomes not a trivial task. SVD helps tackling with this problem by introducing a so-called numerical rank of matrix. Namely, if matrix $\mathbf{U}$ is obtained by experimental measurements, knowing that the accuracy of measurements is, say ± 0.01, then the numerical rank of matrix $\mathbf{U}$ is equal to r if $\sigma_{r+1} < 0.01$. In other words, singular value indicates how close given matrix is to a matrix with the lower rank. Using this feature, SVD can also be used to construct low-order approximation of matrix $\mathbf{U}$.

Since both $\mathbf{V}^1$ and $\mathbf{V}^2$ are orthogonal matrices it is evident from (3.55) that matrix $\mathbf{U}$ can be expressed as

$$\mathbf{U} = \mathbf{V}^1\mathbf{S}\left(\mathbf{V}^2\right)^T \tag{3.58}$$

Now let us suppose that we would like to find matrix with lower rank to approximate matrix $\mathbf{U}$. By keeping just the first K singular values into a square $K \times K$ diagonal matrix $\mathbf{S}_K$ and collecting the corresponding columns from $\mathbf{V}^1$ and $\mathbf{V}^2$, the K^{TH} approximation of $\mathbf{U}$ is given by

$$\mathbf{U} \approx \mathbf{V}_K^1\mathbf{S}_K\left(\mathbf{V}_K^2\right)^T \tag{3.59}$$

The optimality of this approximation lays in the fact that no other rank K matrix can be closer to $\mathbf{U}$ in the least square sense (square root of the sums of squares of all the elements). This is practically the same thing as to seek for K-dimensional subspace for which the mean square distance of the points, from the subspace, is minimized, which indicates the connection between SVD and previously discussed methods for derivation of POD basis.

Unlike orthogonal decomposition, SVD can be applied on rectangular matrices. However, there is a strong connection between the two decompositions. This can be seen by pre-multiplying (3.58) by $\mathbf{U}^T$, having in mind orthogonality of $\mathbf{V}^1$ it will read

$$\mathbf{U}^T\mathbf{U} = \left(\mathbf{V}^1\mathbf{S}\left(\mathbf{V}^2\right)^T\right)^T\mathbf{V}^1\mathbf{S}\left(\mathbf{V}^2\right)^T = \mathbf{V}^2\left(\mathbf{V}^1\mathbf{S}\right)^T\mathbf{V}^1\mathbf{S}\left(\mathbf{V}^2\right)^T \tag{3.60}$$

Using notation introduced in section 3.2, namely, $\mathbf{D} = \mathbf{U}^T \cdot \mathbf{U}$, and once again taking into account orthogonality of $\mathbf{V}^1$ we can write

$$\mathbf{D} = \mathbf{V}^2\mathbf{S}^T\left(\mathbf{V}^1\right)^T\mathbf{V}^1\mathbf{S}\left(\mathbf{V}^2\right)^T = \mathbf{V}^2\mathbf{S}^T\mathbf{S}\left(\mathbf{V}^2\right)^T = \mathbf{V}^2\mathbf{S}^2\left(\mathbf{V}^2\right)^T \tag{3.61}$$

Matrix $\mathbf{S}^2 = \mathbf{S}^T\mathbf{S}$ is also a diagonal $N \times M$ matrix except that its elements are squares of singular values, namely with only the elements on main diagonal different from zero and equal to

$$s_{ii}^2 = \sigma_i^2 \tag{3.62}$$

Finally, taking into account orthogonality of matrix $\mathbf{V}^2$ it can be written

$$\mathbf{D}\mathbf{V}^2 = \mathbf{V}^2\mathbf{S}^2 \tag{3.63}$$

The last equation shows, that columns of $\mathbf{V}^2$ are eigenvectors of matrix $\mathbf{D}$, earlier called modified correlation matrix, and that singular values of matrix $\mathbf{U}$ are square roots of eigenvalues of matrix $\mathbf{D}$ (which are equal to eigenvalues of covariance matrix $\mathbf{C}$, as we have already seen).

Multiplying Eq. 3.58 by $\mathbf{U}^T$ from the right side and following the similar procedure it can be shown that the columns of $\mathbf{V}^1$ are eigenvectors of, earlier introduced covariance matrix $\mathbf{C} = \mathbf{U}\mathbf{U}^T$.

Therefore, there is a full equivalence between approximation given by SVD (Eq. 3.59) and the one previously derived based on POD basis, namely

$$\mathbf{U} \approx \mathbf{V}_K^1\mathbf{S}_K\left(\mathbf{V}_K^2\right)^T = \bar{\mathbf{\Phi}}\bar{\mathbf{A}} \tag{3.64}$$

Furthermore, the equivalence between two previously discussed methods for derivation of PCs (the one that starts from covariance matrix, and the one that starts from modified correlation matrix) is embedded in the SVD. Starting from the definition of singular value, and writing (3.57) for the first singular value and corresponding singular vectors, taking into account the connection between singular values and eigenvalues of matrices $\mathbf{D}$ and $\mathbf{C}$ we have

$$\mathbf{U}\mathbf{v}_1^2 = \sigma_1\mathbf{v}_1^1 = \sqrt{\lambda_1}\mathbf{v}_1^1 \tag{3.65}$$

which gives

$$\mathbf{U}\mathbf{v}_1^2\frac{1}{\sqrt{\lambda_1}} = \mathbf{v}_1^1 \tag{3.66}$$

Recalling that we have shown that vector $\mathbf{v}^2$ is eigenvector of matrix $\mathbf{D}$, and vector $\mathbf{v}^1$ is eigenvector of matrix $\mathbf{C}$ we can confirm that the left-hand side of the equation is exactly what we have derived for the first POD direction starting from modified correlation matrix (Eq. 3.53), while the approach based on covariance matrix showed that the first POD direction is actually first eigenvector of matrix $\mathbf{C}$, here given on the right-hand side of the equation.

Example 3.3. Consider the same set of vectors as the one given in Example 3.1. Show that SVD approximation of rank 1 and rank 2 are giving exactly the same results as the one obtained by PCA achieved in examples 3.1 and 3.2

Solution:

Singular value decomposition of **U** *is resulting in the following matrices:*

$$\mathbf{V}^1 = \begin{bmatrix} -0.5325 & 0.8462 & 0.0183 \\ -0.5255 & -0.3135 & -0.7909 \\ -0.6636 & -0.4308 & 0.6116 \end{bmatrix}$$

$$\mathbf{S} = \begin{bmatrix} 12.308 & 0 & 0 & 0 \\ 0 & 1.208 & 0 & 0 \\ 0 & 0 & 0.213 & 0 \end{bmatrix}$$

$$\mathbf{V}^2 = \begin{bmatrix} -0.1399 & 0.0844 & -0.7572 & 0.6325 \\ -0.6566 & 0.6814 & -0.0669 & -0.3162 \\ -0.5701 & -0.7196 & -0.2390 & -0.3162 \\ -0.4735 & -0.1035 & 0.6042 & 0.6325 \end{bmatrix}$$

<u>*Note that* $\mathbf{V}^1$ *matrix corresponds exactly to eigenvectors of matrix* **C** *and* $\mathbf{V}^2$ *correspond exactly to eigvectors of matrix* **D**</u>

Rank one approximation is obtained by taking just the first eigenvectors and corresponding singular value (Eq. 3.64), namely

$$\mathbf{U}_1 \approx \begin{bmatrix} 0.5325 \\ -0.5255 \\ -0.6636 \end{bmatrix} \cdot 12.308 \cdot \begin{bmatrix} -0.1399 & -0.6566 & -0.5701 & -0.4735 \end{bmatrix} =$$

$$= \begin{bmatrix} 0.9167 & 4.3037 & 3.7366 & 3.1034 \\ 0.9046 & 4.2468 & 3.6872 & 3.0625 \\ 1.1424 & 5.3633 & 4.6566 & 3.8676 \end{bmatrix}$$

Rank two approximation is obtained on analogous way, taking into account the first two eigenvectors, namely

(continued)

$$\mathbf{U}_2 \approx \begin{bmatrix} -0.5325 & 0.8462 \\ -0.5255 & -0.3135 \\ -0.6636 & -0.4308 \end{bmatrix} \begin{bmatrix} 12.308 & 0 \\ 0 & 1.208 \end{bmatrix}$$

$$\times \begin{bmatrix} -0.1399 & -0.6566 & -0.5701 & -0.4735 \\ 0.0844 & 0.6814 & -0.7196 & -0.1035 \end{bmatrix}$$

$$= \begin{bmatrix} 1.0029 & 5.0003 & 3.0009 & 2.9976 \\ 0.8726 & 3.9887 & 3.9598 & 3.1016 \\ 1.0985 & 5.0087 & 5.0311 & 3.9214 \end{bmatrix}$$

Comparing the matrices here computed it can be noted that both of them (rank 1 and rank 2 approximations) gave exactly the same results as those previously calculated in examples 3.1 and 3.2 using just the first, and the first two POD directions.

3.4 Approximation of Discrete Fields Using POD

The main topic of this book are inverse analyses in structural context, that, as it was anticipated in Chap. 1, are combining simulations with experiments within an optimization procedure apt to the identification of some unknown parameters (or in general, some other missing data to formulate the forward problem, i.e. boundary conditions, initial conditions, etc.). Therefore, the final goal is to connect previously presented methodology with structural simulations. In what follows it will be shown how on a very effective way POD can be used to construct high accuracy, low-dimensional approximations of discrete fields like those resulting from numerical simulations.

Numerical techniques, like finite element method, used for structural simulations are solving continuum problems on discretized domains, having as resulting responses discrete fields represented by the values in nodes and integration points (e.g. nodal displacements, values of stresses in Gauss quadrature points, etc.). This fact already suggests that the results from these simulations can easily be treated by discrete POD theory.

In order to establish the connection with what was previously discussed, let us introduce the concept of *snapshots* which is a fundamental notion within the POD theory. In general, one snapshot represents some output of a certain system

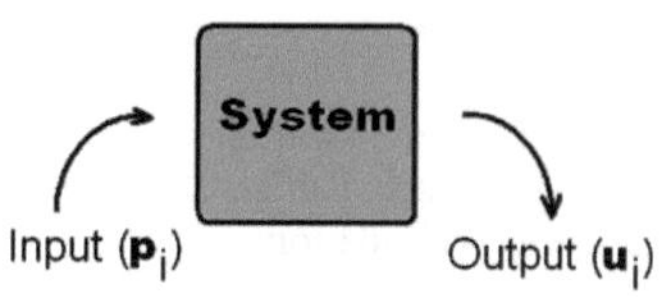

Fig. 3.4 Snapshot as an output of a system

corresponding to some input (Fig. 3.4). In the context presented in this book, in more concrete terms, a snapshot will be a collection of N discrete values of a certain state variable resulting from a simulation (which represents a system) collected in vector $\mathbf{u}_i$, corresponding to some input parameters (collected in vector $\mathbf{p}_i$) on which the solution depends. Therefore, one snapshot can for instance be a vector of nodal displacements that corresponds to a certain combination of material parameters entering into the constitutive model used in the simulation. It should be noted that, in general, the system can also be represented by an experiment, where the snapshot would be a collection of measurements taken from it.

Further, a set of M different snapshots, corresponding to different input parameters, can be collected in a rectangular $N \times M$ matrix $\mathbf{U}$, called the *snapshot matrix*. Therefore, a snapshot matrix represents a collection of responses of one system, under given conditions, corresponding to different values of parameters on which the solution depends. This snapshot matrix can be interpreted as a set of M, N-dimensional vectors. Each vector corresponds to one parameter combination. It is reasonable to expect that there will be a strong correlation between these vectors since they represent the outputs of the same system where just some parameters are changed. The expected correlation suggests that previously presented POD theory can be effectively applied on the snapshot matrix, allowing to construct a new basis in which the dimensionality can be drastically cut-down to $K << N$.

To make the things more concrete, let us imagine that the field of interest is a displacement field of some solid body. The system is then represented by a numerical model of this body, and the resulting nodal displacements from the simulations collected in vectors $\mathbf{u}_i$ are representing snapshots. The inputs to the system, that are changing from one snapshot to another, are some parameters entering into the constitutive model of material, while all the boundary conditions and initial conditions are the same for all of the snapshots. Snapshot matrix can be generated using this numerical model, and the process will involve a set of M simulations, in which the parameters are changing within some specified range while keeping both boundary conditions and loads fixed (Fig. 3.5). All the snapshots are collected in rectangular matrix $\mathbf{U}$, such that different columns are corresponding to discretized displacement fields (represented by nodal displacements) of solid bodies of the same geometry, subjected to the same constraints and loads, built of

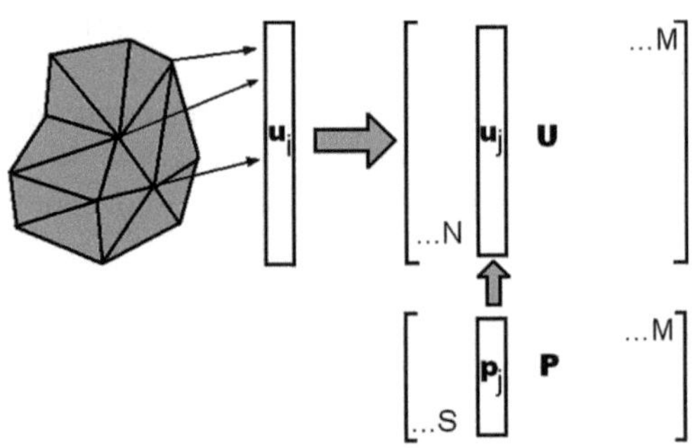

Fig. 3.5 Generation of snapshot matrix using numerical simulations

different materials. In the most trivial case when the analyses are linear (for instance the only parameter that changes form one snapshot to another is Young modulus) there will be a full linear dependence between the snapshots. In such case, it would be possible to express all the snapshots, without any loss of accuracy, in the reduced space of just one dimension.

The process of building the snapshot matrix and further construction of the POD basis will be illustrated on the following structural example.

Example 3.4. Consider the truss structure presented on the figure below.

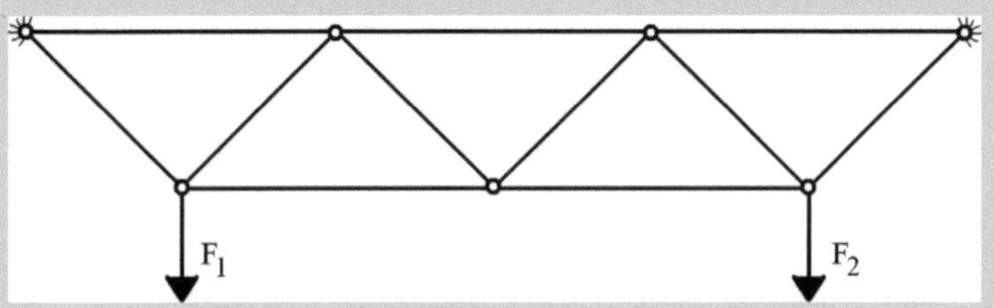

Keeping fixed boundary conditions and forces equal to $F_1 = 5{,}000$ N and $F_2 = 2{,}000$ N, generate the snapshot matrix of nodal displacements for the given structure, by varying the following two parameters: Young's modulus from 125 GPa to 200 GPa, with the step 25 GPa, and area of cross-section of the trusses from 6 mm^2 to 12 mm^2 with the step 2 mm^2. Construct the POD basis of the snapshot matrix and show that in this case the dimensionality can be reduced to 1 without any loss of accuracy.

Solution:

First, we need to build a numerical model for a given structure. For this purpose, a two simple MATLAB codes can be used. The first one is the main MATLAB routine that solves structural problem using truss finite elements, and the second one is a MATLAB function that computes stiffness matrix of single truss element, called by the main routine.

(continued)

```
*************************************************************
% Main program for truss structure (file name truss_str.m)
% Coordinates - units L[mm]
COO=[0,0;2,0;4,0;1,1;3,1;-1,1;5,1];
COO=1000*COO;
DOF=size(COO,1)*2;
% Elements
ELM=[1,2;2,3;6,4;4,5;5,7;1,6;1,4;2,4;2,5;3,5;3,7];
% Properties (units F[N], L[mm])
E=125000; A=6;
% Assembling of stiffness matrix
MSTF(1:DOF,1:DOF)=0;
for i=1:size(ELM,1)
    X(1,:)=COO(ELM(i,1),:);
    X(2,:)=COO(ELM(i,2),:);
    STF=trstiff(X,E,A);
    MSTF(2*ELM(i,1)-1:2*ELM(i,1),2*ELM(i,1)-
1:2*ELM(i,1))=MSTF(2*ELM(i,1)-1:2*ELM(i,1),2*ELM(i,1)-
1:2*ELM(i,1))+STF(1:2,1:2);
    MSTF(2*ELM(i,1)-1:2*ELM(i,1),2*ELM(i,2)-
1:2*ELM(i,2))=MSTF(2*ELM(i,1)-1:2*ELM(i,1),2*ELM(i,2)-
1:2*ELM(i,2))+STF(3:4,1:2);
    MSTF(2*ELM(i,2)-1:2*ELM(i,2),2*ELM(i,1)-
1:2*ELM(i,1))=MSTF(2*ELM(i,2)-1:2*ELM(i,2),2*ELM(i,1)-
1:2*ELM(i,1))+STF(1:2,3:4);
    MSTF(2*ELM(i,2)-1:2*ELM(i,2),2*ELM(i,2)-
1:2*ELM(i,2))=MSTF(2*ELM(i,2)-1:2*ELM(i,2),2*ELM(i,2)-
1:2*ELM(i,2))+STF(3:4,3:4);
end
% Constrains
con=[12,13,14,15]; % constrained DOF
CDOF=size(con,2);
% Forces;
dF(DOF,1)=0;dF(2)=-5000;dF(6)=-2000;
% Reduced stiffness matrix and force vector
RSTF=MSTF(1:DOF-CDOF,1:DOF-CDOF);
dFR=dF(1:DOF-CDOF);
% Solving for displacements
d=inv(RSTF)*dFR;
% End of main program
*************************************************************
```

```
*************************************************************
function STF=trstiff(coo,E,A)
% Function that computes stiffness matrix for truss element
% (file name trstiff.m)
L=sqrt((coo(2,1)-coo(1,1))^2+(coo(2,2)-coo(1,2))^2);
C=(coo(2,1)-coo(1,1))/L;
S=(coo(2,2)-coo(1,2))/L;
% Stiffness matrix
STF=E*A/L*[C^2,S*C,-C^2,-S*C;
           S*C,S^2,-S*C,-S^2;
           -C^2,-S*C,C^2,S*C;
           -S*C,-S^2,S*C,S^2];
*************************************************************
```

(*continued*)

Note that the program uses reduced stiffness matrix, and therefore as a result gives the displacements of just those nodes that are not constrained, having therefore the displacement vector of the dimension 10 × 1

In the listing of program the parameters are specified to coincide with the first values from the ranges for E and A. The program should be run for each of 16 pairs of values for the parameters within the specified range. The resulting displacement vector should be collected in the snapshot matrix. Note that this part doesn't have to be done manually, but within another MATLAB routine with the listing given below (and by ereasing two lines in the main program in which the values are prescribed to the parameters E and A, since here this is done outside of the main program.

```
*****************************************************************
cnt=0;
for E=125000:25000:200000
    for A=6:2:12
        cnt=cnt+1;
        truss_str;
        U(:,cnt)=d;
    end
end
*****************************************************************
```

The resulting snapshot matrix is collecting responses of these 16 different structures:

$$
\mathbf{U}=\left[\begin{array}{ccccccccc}
-11.56 & -8.67 & -6.93 & -5.78 & -9.63 & -7.22 & -5.78 & -4.81 & -8.25 \\
-28.53 & -21.39 & -17.12 & -14.26 & -23.77 & -17.83 & -14.26 & -11.89 & -20.38 \\
-0.89 & -0.67 & -0.53 & -0.44 & -0.74 & -0.56 & -0.44 & -0.37 & -0.63 \\
-31.87 & -23.90 & -19.12 & -15.93 & -26.55 & -19.92 & -15.93 & -13.28 & -22.76 \\
7.11 & 5.33 & 4.27 & 3.56 & 5.93 & 4.44 & 3.56 & 2.96 & 5.08 \\
-16.54 & -12.40 & -9.92 & -8.27 & -13.78 & -10.34 & -8.27 & -6.89 & -11.81 \\
-2.67 & -2.00 & -1.60 & 1.33 & -2.22 & -1.67 & -1.33 & -1.11 & -1.90 \\
-35.53 & -26.65 & -21.32 & -17.76 & -29.61 & -22.21 & -17.76 & -14.80 & -25.38 \\
-2.67 & -2.00 & -1.60 & -1.33 & -2.22 & -1.67 & -1.33 & -1.11 & -1.90 \\
-28.20 & -21.15 & -16.92 & -14.10 & -23.50 & -17.63 & -14.10 & -11.75 & -20.14
\end{array}\right.
$$

$$
\left.\begin{array}{ccccccc}
-6.19 & -4.95 & -4.13 & -7.22 & -5.42 & -4.33 & -3.61 \\
-15.28 & -12.23 & -10.19 & -17.83 & -13.37 & -10.70 & -8.91 \\
-0.48 & -0.38 & -0.32 & -0.56 & -0.42 & -0.33 & -0.28 \\
-17.07 & -13.66 & -11.38 & -19.92 & -14.94 & -11.95 & -9.96 \\
3.81 & 3.05 & 2.54 & 4.44 & 3.33 & 2.67 & 2.22 \\
-8.86 & -7.09 & -5.91 & -10.34 & -7.75 & -6.20 & -5.17 \\
-1.43 & -1.14 & -0.95 & -1.67 & -1.25 & -1.00 & -0.83 \\
-19.03 & -15.23 & -12.69 & -22.21 & -16.65 & -13.32 & -11.10 \\
-1.43 & -1.14 & -0.95 & -1.67 & -1.25 & -1.00 & -0.83 \\
-15.11 & -12.09 & -10.07 & -17.63 & -13.22 & -10.58 & -8.81
\end{array}\right]
$$

(*continued*)

Now we should construct POD basis for the snapshot matrix. We can use first approach, since matrix **C** *in this case is 10×10, while matrix* **D** *is 16×16. Solving eigenvalue problem in MATLAB is possible by the use of already implemented command, namely*

```
[V,L]=eigs(U*U',10)
```

which as a result gives 10×10 matrix **V**, *whose columns are eigenvectors of matrix* **C** *($\mathbf{C} = \mathbf{U}\mathbf{U}^T$) and 10×10 matrix* **L**, *whit only diagonal elements different from zero and equal to eigenvalues of matrix* **C**.

It may be observed that only the first eigenvalue of matrix **C** *has finite value and is equal to 2.458E + 4, while already second one is equal to 4.663E-12, which is a numerical zero. This proves that the responses of the system are fully correlated and therefore, they can be interpreted without any loss of accuracy in the new one-dimensional basis. The vector of this new basis coincides with the first eigenvector of matrix* **C**, *and is equal to:*

$$\bar{\Phi} = \begin{bmatrix} 0.175 & 0.013 & 0.483 & -0.108 & 0.250 & 0.040 & 0.538 & 0.040 & 0.427 \end{bmatrix}^T$$

The amplitudes in the reduced space are computed by

$$\bar{\mathbf{A}} = \bar{\Phi}^T \mathbf{U} = \begin{bmatrix} -66.027 & -49.520 & -39.616 & -33.013 & -55.022 & -41.267 & -33.013 \\ -27.511 & -47.162 & -35.372 & -28.297 & -23.581 & -41.267 & -30.950 & -24.760 & -20.633 \end{bmatrix}$$

With this, the dimensionality of the problem was cut-down from 10 to just 1. In the new basis, the response (i.e. snapshot) of each of 16 previous structures (defined by different values of E and A) is given by just one number (corresponding amplitude from vector $\bar{\mathbf{A}}$*). Since the correlation was full due to the linearity of the problem, no error was introduced by this truncation. To reconstruct back the response of any of the snapshots it is enough to make a simple multiplication between reduced basis and corresponding amplitude, for instance nodal displacements of first structure are equal to*

$$\mathbf{u} = -66.027 \begin{bmatrix} 0.175 \\ 0.432 \\ 0.013 \\ 0.483 \\ -0.108 \\ 0.250 \\ 0.040 \\ 0.538 \\ 0.040 \\ 0.427 \end{bmatrix} = \begin{bmatrix} -11.56 \\ -28.53 \\ -0.89 \\ -31.87 \\ 7.11 \\ -16.54 \\ -2.67 \\ -35.53 \\ -2.67 \\ -28.20 \end{bmatrix}$$

(continued)

Comparing the result with the original snapshot matrix, it is evident that the displacements are completely the same. This can easily be verified for any other snapshot.

Last exercise showed how the discrete POD theory can be used to construct low dimensional approximation of structural problems. In the linear example treated here, due to the full correlation between the responses, the dimensionality was reduced to just one component. However, this formulation is still not very useful for the substitution of simulations since actually what we did up to now is just expressed the existing responses of the system (previously computed by "full" numerical model) in new basis where they can be, in this particular case, represented by just one number. But this "light" model obviously can give the responses of the system for just a discrete number of parameter combinations, those that were previously used to generate the snapshot matrix. Since within inverse analyses, a multiple system responses computations are usually performed, therefore a required formulation needs to be capable of computing an approximate response of the system for any arbitrary combination of parameters (like finite element simulations are doing).

This can be done following the procedure, first introduced by Bialecki et al. in [7] and [16], in thermal problems, and then later applied in structural problems by Buljak et al. in [17–19]. This technique combines POD with Radial Basis Functions (RBF) to interpolate previously generated results. Before coming to details on how the method is implemented in the present context, let us first see how RBF are used for scattered data interpolation.

3.5 Radial Basis Functions for Scattered Data Interpolation

Radial Basis Functions (RBF) are frequently used when it is needed to construct an approximation of a certain multivariable function by interpolation between the existing data. This section will give a very brief description of the method to make the reader familiar with main principles and to show how it can be effectively combined with previously described POD theory in the structural context here of interest. For more detailed description on RBF reader should refer to [20].

Let us assume that we wish to approximate some function $f(\mathbf{x})$, where $\mathbf{x}$ is an M-dimensional vector, for which the only information that we have is a set of, say N values $\mathbf{x}_i$, called "nodes", for which the values of the function are known. In classical, local methods of interpolation, the problem is solved locally in the neighborhood of the point $\mathbf{x}$ for which the value of the function is required. In this case it doesn't exist just one continuous function defined over the whole domain where the data are situated. Instead, for any arbitrary value of $\mathbf{x}$ the interpolation is performed involving just a few nearby data.

The approach of RBF is different, and it seeks for one continuous function defined over the whole domain and that depends on the entire data set defined by the pairs of given N nodes and their values of the function. Therefore, the approximation of the function is written as a linear combination of some functions g_i, that in general case can be non-linear functions, namely

$$f(\mathbf{x}) \approx \sum_{i=1}^{N} \alpha_i \cdot g_i(\mathbf{x}) \tag{3.67}$$

where α_i are coefficients of this combination. The last equation is completely defined once the basis functions g_i are selected and the coefficients α_i are known. Choosing a set of any radial basis functions for the basis, like for instance Euclidian distance

$$g_i(\mathbf{x}) = g(\|\mathbf{x} - \mathbf{x}_i\|), i = 1, 2, ..., N \tag{3.68}$$

all that remains is to determine coefficients α_i. This is done from the known N values of the function in the nodes $\mathbf{x}_i$. The condition from which unknown coefficients are determined states, that the interpolation needs to be exact in all N nodes, therefore, the approximation (3.67) needs to be written for all N nodes which gives a system of N linear equations defined by

$$f(\mathbf{x}_j) = y_j = \sum_{i=1}^{N} \alpha_i \cdot g_i(\mathbf{x}_j), j = 1, ..., N \tag{3.69}$$

where y_j are known values of the function in the nodes. Introducing the following matrix notation

$$\begin{gathered}\mathbf{G} = \begin{bmatrix} g_1(\mathbf{x}_1) & \dots & g_N(\mathbf{x}_1) \\ \dots & \dots & \dots \\ g_1(\mathbf{x}_N) & \dots & g_N(\mathbf{x}_N) \end{bmatrix}; \\ \boldsymbol{\alpha} = [\ \alpha_1,\ \alpha_2, ...,\ \alpha_\mathrm{n}]^T; \mathbf{Y} = [y_1, y_2, ..., y_N]^T\end{gathered} \tag{3.70}$$

the system (3.69) can be written as

$$\boldsymbol{\alpha} \cdot \mathbf{G} = \mathbf{Y} \tag{3.71}$$

Matrix equation (3.71) can be solved for unknown interpolation coefficients α_i. After this, it is possible by the use of (3.67) to obtain the approximation of the function in any given point. Considering the way how the interpolation coefficients are determined, (3.67) will give exact results in the nodes, and some form of interpolated value for any other value of $\mathbf{x}$. The interpolation coefficients are computed once for all, and they obviously involve all known values of functions which represents already anticipated difference with respect to the classical, local interpolation techniques. Therefore RBF through Eq. 3.67 are giving one approximation valid for the whole domain where original data were situated. Another advantage of RBF is that the distribution of nodes in their space needs not to be regular like in some other interpolations (for instance Lagrangian interpolation [21]), but it can also be scattered. Of course the distribution of nodes is influencing the error of interpolation which can be easier kept under control when a regular grid is used, it is important to note that this is not a necessary condition for the method.

Apart of the number of nodes and their distribution, the error of interpolation is influenced also by the choice of RBF. The most frequently used examples are listed below [22]

$$Thin-platesplines,\ \left\|\mathbf{x}-\mathbf{x}_j\right\| \ln\left(\left\|\mathbf{x}-\mathbf{x}_j\right\|\right) \tag{3.72}$$

$$Linearsplines,\ \left\|\mathbf{x}-\mathbf{x}_j\right\| \tag{3.73}$$

$$Cubicspline,\ \left\|\mathbf{x}-\mathbf{x}_j\right\|^3 \tag{3.74}$$

$$Gaussian,\ \exp\left(-\frac{\left\|\mathbf{x}-\mathbf{x}_j\right\|}{c_j^2}\right) \tag{3.75}$$

$$Multiquadric\sqrt{1+\frac{\left\|\mathbf{x}-\mathbf{x}_j\right\|^2}{c_j^2}} \tag{3.76}$$

Previous equations are written for the case when the value of the function is a scalar. However, the interpolation can also be performed when the value of the functions is a vector, say S-dimensional one. In this case (3.69) becomes

$$\begin{bmatrix}\alpha_{11}\\ \alpha_{21}\\ \dots\\ \alpha_{S1}\end{bmatrix}\cdot g_1(\mathbf{x}_i)+\begin{bmatrix}\alpha_{12}\\ \alpha_{22}\\ \dots\\ \alpha_{S2}\end{bmatrix}\cdot g_2(\mathbf{x}_i)+\dots+\begin{bmatrix}\alpha_{1N}\\ \alpha_{2N}\\ \dots\\ \alpha_{SN}\end{bmatrix}\cdot g_N(\mathbf{x}_i)=\begin{bmatrix}y_1^i\\ y_2^i\\ \dots\\ y_S^i\end{bmatrix} \tag{3.77}$$

As it may be observed, here the coefficients of interpolation are forming $S \times N$ matrix instead of vector like in previous case. Equation 3.77, for one pair of $\mathbf{x}_i$ and $\mathbf{y}_i$, can be written as S algebraic equations in the following form

$$\begin{gathered}\alpha_{j1}\cdot g_1(\mathbf{x}_i)+\ \alpha_{j2}\cdot g_2(\mathbf{x}_i)+\dots+\ \alpha_{jN}\cdot g_N(\mathbf{x}_i)+\ \alpha_{jN+1}=y_j^i\\ i=1,2,\dots N; j=1,2,\dots,S\end{gathered} \tag{3.78}$$

Equation 3.78 is practically the same one as (3.69), except that here it is written for each component of the vector $\mathbf{y}_i$. Writing (3.78) for each N pairs of $\mathbf{x}_i$ and $\mathbf{y}_i$ a system of $N \times S$ equations is obtained that should be solved for the unknown interpolation coefficients α_{ij}. Let us use the generic formula given by (3.78) for the simple case where $S = 2$ and $N = 3$, just to show how the system is transformed to a matrix equation. Using notation given by (3.77) and (3.78) the resulting system of algebraic equations is given by

$$\begin{aligned}
\alpha_{11}g_1(\mathbf{x}_1) + \alpha_{12}g_2(\mathbf{x}_1) + \alpha_{13}g_2(\mathbf{x}_1) &= y_1^1 \\
\alpha_{21}g_1(\mathbf{x}_1) + \alpha_{22}g_2(\mathbf{x}_1) + \alpha_{23}g_2(\mathbf{x}_1) &= y_2^1 \\
\alpha_{11}g_1(\mathbf{x}_2) + \alpha_{12}g_2(\mathbf{x}_2) + \alpha_{13}g_2(\mathbf{x}_2) &= y_1^2 \\
\alpha_{21}g_1(\mathbf{x}_2) + \alpha_{22}g_2(\mathbf{x}_2) + \alpha_{23}g_2(\mathbf{x}_2) &= y_2^2 \\
\alpha_{11}g_1(\mathbf{x}_3) + \alpha_{12}g_2(\mathbf{x}_3) + \alpha_{13}g_2(\mathbf{x}_3) &= y_1^3 \\
\alpha_{21}g_1(\mathbf{x}_3) + \alpha_{22}g_2(\mathbf{x}_3) + \alpha_{23}g_2(\mathbf{x}_3) &= y_2^3
\end{aligned} \tag{3.79}$$

First two equations are corresponding to the first pair of $\mathbf{x}$ and $f(\mathbf{x}) = \mathbf{y}$, the second two correspond to the second pair, and the last two to the last pair. Writing the system (3.79) in a matrix form it will become

$$\begin{bmatrix}
g_1(\mathbf{x}_1) & g_2(\mathbf{x}_1) & g_3(\mathbf{x}_1) & 0 & 0 & 0 \\
0 & 0 & 0 & g_1(\mathbf{x}_1) & g_2(\mathbf{x}_1) & g_3(\mathbf{x}_1) \\
g_1(\mathbf{x}_2) & g_2(\mathbf{x}_2) & g_3(\mathbf{x}_2) & 0 & 0 & 0 \\
0 & 0 & 0 & g_1(\mathbf{x}_2) & g_2(\mathbf{x}_2) & g_3(\mathbf{x}_2) \\
g_1(\mathbf{x}_3) & g_2(\mathbf{x}_3) & g_3(\mathbf{x}_3) & 0 & 0 & 0 \\
0 & 0 & 0 & g_1(\mathbf{x}_3) & g_2(\mathbf{x}_3) & g_3(\mathbf{x}_3)
\end{bmatrix} \cdot \begin{bmatrix} \alpha_{11} \\ \alpha_{12} \\ \alpha_{13} \\ \alpha_{21} \\ \alpha_{22} \\ \alpha_{23} \end{bmatrix}
= \begin{bmatrix} y_1^1 & y_1^2 & y_1^3 \\ y_2^1 & y_2^2 & y_2^3 \end{bmatrix} \tag{3.80}$$

After regrouping the last equation can be written as

$$\begin{bmatrix} \alpha_{11} & \alpha_{12} & \alpha_{13} \\ \alpha_{21} & \alpha_{22} & \alpha_{23} \end{bmatrix} \cdot \begin{bmatrix} g_1(\mathbf{x}_1) & g_1(\mathbf{x}_2) & g_1(\mathbf{x}_3) \\ g_2(\mathbf{x}_1) & g_2(\mathbf{x}_2) & g_2(\mathbf{x}_3) \\ g_3(\mathbf{x}_1) & g_3(\mathbf{x}_2) & g_3(\mathbf{x}_3) \end{bmatrix} = \begin{bmatrix} y_1^1 & y_1^2 & y_1^3 \\ y_2^1 & y_2^2 & y_2^3 \end{bmatrix} \tag{3.81}$$

Denoting by $\mathbf{A}$ the matrix with the coefficients and matrix with the values of the function by $\mathbf{Y}$ we can write

$$\mathbf{A} \cdot \mathbf{G} = \mathbf{Y} \tag{3.82}$$

which is identical to (3.71) except that here $\mathbf{A}$ and $\mathbf{Y}$ are matrices and not vectors. The last equation is solved for the unknown matrix $\mathbf{A}$.

Once the coefficients are determined, the interpolation is obtained by the following equation

$$f(\mathbf{x}) \approx \begin{bmatrix} \alpha_{11} & \alpha_{12} & \cdots & \alpha_{1N} \\ \alpha_{21} & \alpha_{22} & \cdots & \alpha_{2N} \\ \cdots & \cdots & \cdots & \cdots \\ \alpha_{S1} & \alpha_{S2} & \cdots & \alpha_{SN} \end{bmatrix} \cdot \begin{bmatrix} g_1(\mathbf{x}) \\ g_2(\mathbf{x}) \\ \cdots \\ g_N(\mathbf{x}) \end{bmatrix} \tag{3.83}$$

It should be noted that the number of nodes N can be in general case relatively large, involving therefore a large system of linear equations to be solved in order to determine interpolation coefficients. However, this process is done once-for-all, and after that, the interpolation is performed directly by matrix multiplication given by (3.83).

Let us consider a following simple numerical example where the function is given in analytical form. This will allow us to have a direct comparison between interpolated values and computed ones in order to see the error of interpolation.

In this example we saw that there is a considerable influence of the choice of the type of RBF to the error of interpolation. However, there is no single rule to use as a guideline for this choice and in practical examples it will depend on the case.

Example 3.5. Argument of the function is defined as two-component vector $\mathbf{x} = \begin{bmatrix} x_1 & x_2 \end{bmatrix}^T$, while value of the function is given as three-component vector defined by $\mathbf{y} = \begin{bmatrix} \sqrt{x_1} & x_2^2 & x_1 + x_2 \end{bmatrix}^T$. Use as nodes values of the function in the following nine points:

$$\mathbf{X} = \begin{bmatrix} 1 & 2 & 3 & 1 & 2 & 3 & 1 & 2 & 3 \\ 1 & 1 & 1 & 2 & 2 & 2 & 3 & 3 & 3 \end{bmatrix}$$

which are forming regular grid over domain $[1, 3]$ for both components. Build RBF interpolation for the following three points

$$\begin{bmatrix} 1.5 & 1.1 & 2.5 \\ 1.5 & 1.2 & 2.5 \end{bmatrix}$$

Compare the results using linear spline (3.73) and cubic spline (3.74) as a basis functions. Compare both interpolations with the analytical solution.

Solution:

First, we should compute the solution in all 9 nodes and collect the results in matrix $\mathbf{Y}$

$$\mathbf{Y} = \begin{bmatrix} 1.00 & 1.41 & 1.73 & 1.00 & 1.41 & 1.73 & 1.00 & 1.41 & 1.73 \\ 1.00 & 1.00 & 1.00 & 4.00 & 4.00 & 4.00 & 9.00 & 9.00 & 9.00 \\ 2.00 & 3.00 & 4.00 & 3.00 & 4.00 & 5.00 & 4.00 & 5.00 & 6.00 \end{bmatrix}$$

(*continued*)

Second, we should compute matrices $\mathbf{G}_1$ *and* $\mathbf{G}_2$ *that are using RBF of the types given by (3.73) and (3.74) respectively. In this case, this will be 9×9 matrices, since there is total of 9 nodes for which the values of the function are known.*

$$\mathbf{G}_1 = \begin{bmatrix} 0.00 & 1.00 & 2.00 & 1.00 & 1.41 & 2.24 & 2.00 & 2.24 & 2.83 \\ 1.00 & 0.00 & 1.00 & 1.41 & 1.00 & 1.41 & 2.24 & 2.00 & 2.24 \\ 2.00 & 1.00 & 0.00 & 2.24 & 1.41 & 1.00 & 2.83 & 2.24 & 2.00 \\ 1.00 & 1.41 & 2.24 & 0.00 & 1.00 & 2.00 & 1.00 & 1.41 & 2.24 \\ 1.41 & 1.00 & 1.41 & 1.00 & 0.00 & 1.00 & 1.41 & 1.00 & 1.41 \\ 2.24 & 1.41 & 1.00 & 2.00 & 1.00 & 0.00 & 2.24 & 1.41 & 1.00 \\ 2.00 & 2.24 & 2.83 & 1.00 & 1.41 & 2.24 & 0.00 & 1.00 & 2.00 \\ 2.24 & 2.00 & 2.24 & 1.41 & 1.00 & 1.41 & 1.00 & 0.00 & 1.00 \\ 2.83 & 2.24 & 2.00 & 2.24 & 1.41 & 1.00 & 2.00 & 1.00 & 0.00 \end{bmatrix}$$

$$\mathbf{G}_2 = \begin{bmatrix} 0.00 & 1.00 & 8.00 & 1.00 & 2.83 & 11.18 & 8.00 & 11.18 & 22.63 \\ 1.00 & 0.00 & 1.00 & 2.83 & 1.00 & 2.83 & 11.18 & 8.00 & 11.18 \\ 8.00 & 1.00 & 0.00 & 11.18 & 2.83 & 1.00 & 22.63 & 11.18 & 8.00 \\ 1.00 & 2.83 & 11.18 & 0.00 & 1.00 & 8.00 & 1.00 & 2.83 & 11.18 \\ 2.83 & 1.00 & 2.83 & 1.00 & 0.00 & 1.00 & 2.83 & 1.00 & 2.83 \\ 11.18 & 2.83 & 1.00 & 8.00 & 1.00 & 0.00 & 11.18 & 2.83 & 1.00 \\ 8.00 & 11.18 & 22.63 & 1.00 & 2.83 & 11.18 & 0.00 & 1.00 & 8.00 \\ 11.18 & 8.00 & 11.18 & 2.83 & 1.00 & 2.83 & 1.00 & 0.00 & 1.00 \\ 22.63 & 11.18 & 8.00 & 11.18 & 2.83 & 1.00 & 8.00 & 1.00 & 0.00 \end{bmatrix}$$

Note that matrices $\mathbf{G}_1$ *and* $\mathbf{G}_2$ *are symmetric. This is always the case, since* $g_i(\mathbf{x}_j) = \|\mathbf{x}_i - \mathbf{x}_j\| = g_j(\mathbf{x}_i) = \|\mathbf{x}_j - \mathbf{x}_i\|$

Next, we should solve Eq. 3.82 *for matrix* $\mathbf{A}$*, namely*

$$\mathbf{AG} = \mathbf{Y} \Rightarrow \mathbf{G}^T\mathbf{A}^T = \mathbf{Y}^T \Rightarrow \mathbf{A} = \left(\left(\mathbf{G}^T\right)^{-1}\mathbf{Y}^T\right)^T = \mathbf{Y}\left(\left(\mathbf{G}^T\right)^{-1}\right)^T$$

which is here resulting with two matrices with interpolation coefficients corresponding to two different types of RBF

$$\mathbf{A}_1 = \begin{bmatrix} 0.37 & -0.04 & 0.15 & 0.06 & -0.22 & -0.03 & 0.37 & -0.04 & 0.15 \\ 1.73 & 0.16 & 1.73 & 0.82 & -0.13 & 0.82 & -0.62 & -0.93 & -0.62 \\ 1.30 & 0.13 & 0.71 & 0.13 & -0.57 & -0.15 & 0.71 & -0.15 & 0.13 \end{bmatrix}$$

$$\mathbf{A}_2 = \begin{bmatrix} 0.28 & -0.33 & 0.21 & -0.38 & 0.07 & -0.32 & 0.28 & -0.33 & 0.21 \\ 0.98 & -0.97 & 0.98 & -1.05 & 0.10 & -1.05 & 0.32 & -0.28 & 0.32 \\ 0.86 & -1.04 & 0.69 & -1.04 & 0.19 & -0.86 & 0.69 & -0.86 & 0.52 \end{bmatrix}$$

(*continued*)

Now, for each of the three required points interpolation is performed using (3.83), which means that a vector $\mathbf{g}$ *with values of RBF for the particular point needs to be computed. Denoting by* $\mathbf{g}_1$ *vector of RBF given by (3.73)*

$$\mathbf{g}_1(\mathbf{x}) = \begin{bmatrix} \|\mathbf{x}-\mathbf{x}_1\| \\ \|\mathbf{x}-\mathbf{x}_2\| \\ \|\mathbf{x}-\mathbf{x}_3\| \\ \|\mathbf{x}-\mathbf{x}_4\| \\ \|\mathbf{x}-\mathbf{x}_5\| \\ \|\mathbf{x}-\mathbf{x}_6\| \\ \|\mathbf{x}-\mathbf{x}_7\| \\ \|\mathbf{x}-\mathbf{x}_8\| \\ \|\mathbf{x}-\mathbf{x}_9\| \end{bmatrix} \Rightarrow \mathbf{g}_1^1\left(\begin{bmatrix}1.5\\1.5\end{bmatrix}\right) = \begin{bmatrix}0.71\\0.71\\1.58\\0.71\\0.71\\1.58\\1.58\\1.58\\2.12\end{bmatrix},$$

$$\mathbf{g}_1^2\left(\begin{bmatrix}1.1\\1.2\end{bmatrix}\right) = \begin{bmatrix}0.22\\0.92\\1.91\\0.81\\1.20\\2.06\\1.80\\2.01\\2.62\end{bmatrix}, \mathbf{g}_1^3\left(\begin{bmatrix}2.5\\2.5\end{bmatrix}\right) = \begin{bmatrix}2.12\\1.58\\1.58\\1.58\\0.71\\0.71\\1.58\\0.71\\0.71\end{bmatrix}$$

Interpolated result for the first point is obtained by matrix multiplication:

$$f\left(\begin{bmatrix}1.5\\1.5\end{bmatrix}\right) \approx \begin{bmatrix} 0.37 & -0.04 & 0.15 & 0.06 & -0.22 & -0.03 & 0.37 & -0.04 & 0.15 \\ 1.73 & 0.16 & 1.73 & 0.82 & -0.13 & 0.82 & -0.62 & -0.93 & -0.62 \\ 1.30 & 0.13 & 0.71 & 0.13 & -0.57 & -0.15 & 0.71 & -0.15 & 0.13 \end{bmatrix} \cdot \begin{bmatrix}0.71\\0.71\\1.58\\0.71\\0.71\\1.58\\1.58\\1.58\\2.12\end{bmatrix}$$

$$= \begin{bmatrix}1.138\\2.077\\2.757\end{bmatrix}$$

(*continued*)

and for the remaining two points using the corresponding vectors $\mathbf{g}$, *we have*

$$f\left(\begin{bmatrix}1.1\\1.2\end{bmatrix}\right) \approx \begin{bmatrix}1.018\\1.407\\2.204\end{bmatrix};\ f\left(\begin{bmatrix}2.5\\2.5\end{bmatrix}\right) \approx \begin{bmatrix}1.530\\6.357\\4.897\end{bmatrix}$$

Further, values of $\mathbf{g}_2$ *vectors for RBF given by (3.74) for the three points of interest are*

$$\mathbf{g}_2(\mathbf{x}) = \begin{bmatrix}\|\mathbf{x}-\mathbf{x}_1\|^3\\\|\mathbf{x}-\mathbf{x}_2\|^3\\\|\mathbf{x}-\mathbf{x}_3\|^3\\\|\mathbf{x}-\mathbf{x}_4\|^3\\\|\mathbf{x}-\mathbf{x}_5\|^3\\\|\mathbf{x}-\mathbf{x}_6\|^3\\\|\mathbf{x}-\mathbf{x}_7\|^3\\\|\mathbf{x}-\mathbf{x}_8\|^3\\\|\mathbf{x}-\mathbf{x}_9\|^3\end{bmatrix} \Rightarrow \mathbf{g}_2^1\left(\begin{bmatrix}1.5\\1.5\end{bmatrix}\right) = \begin{bmatrix}0.35\\0.35\\3.95\\0.35\\0.35\\3.95\\3.95\\3.95\\9.54\end{bmatrix},$$

$$\mathbf{g}_2^2\left(\begin{bmatrix}1.1\\1.2\end{bmatrix}\right) = \begin{bmatrix}0.01\\0.78\\6.97\\0.52\\1.75\\8.76\\5.86\\8.15\\17.93\end{bmatrix},\ \mathbf{g}_2^3\left(\begin{bmatrix}2.5\\2.5\end{bmatrix}\right) = \begin{bmatrix}9.54\\3.95\\3.95\\3.95\\0.35\\0.35\\3.95\\0.35\\0.35\end{bmatrix}$$

Interpolated results are obtained by multiplying matrix $\mathbf{A}_2$ *with the corresponding vector* $\mathbf{g}_2$, *namely:*

$$f\left(\begin{bmatrix}1.5\\1.5\end{bmatrix}\right) \approx \begin{bmatrix}1.292\\2.584\\3.251\end{bmatrix};\ f\left(\begin{bmatrix}1.1\\1.2\end{bmatrix}\right) \approx \begin{bmatrix}1.137\\1.811\\2.593\end{bmatrix};\ f\left(\begin{bmatrix}2.5\\2.5\end{bmatrix}\right) \approx \begin{bmatrix}1.623\\6.205\\5.061\end{bmatrix}$$

Finally, the exact values of the function for the three points are given in the following matrix

$$\left[f\left(\begin{bmatrix}1.5\\1.5\end{bmatrix}\right)\ f\left(\begin{bmatrix}1.1\\1.2\end{bmatrix}\right)\ f\left(\begin{bmatrix}2.5\\2.5\end{bmatrix}\right)\right] = \begin{bmatrix}1.225 & 1.049 & 1.581\\2.250 & 1.440 & 6.250\\3.000 & 2.300 & 5.000\end{bmatrix}$$

Generally, one should try couple of functions and compare the error in order to choose the one that works the best for the given example. RBF approximation, as previously mentioned, gives exact results in the nodes for which the data existed. This can be simple proved on the previous numerical example, using computed matrices in one of the nodal points. This also suggests that the error of interpolation is influenced by how far the desired point is situated from some of the nodal points. This feature was also evidenced in the example and can be seen by comparing the error for the first two points. For example the differences between interpolated value and analytical one in the case of linear spline RBF and the analytical one for the first two points are equal to

$$f\left(\begin{bmatrix}1.5\\1.5\end{bmatrix}\right) - \mathbf{y}_1 = \begin{bmatrix}1.138\\2.077\\2.757\end{bmatrix} - \begin{bmatrix}1.225\\2.250\\3.000\end{bmatrix} = \begin{bmatrix}-0.086\\-0.173\\-0.243\end{bmatrix}$$

$$f\left(\begin{bmatrix}1.1\\1.2\end{bmatrix}\right) - \mathbf{y}_2 = \begin{bmatrix}1.018\\1.407\\2.204\end{bmatrix} - \begin{bmatrix}1.049\\1.440\\2.300\end{bmatrix} = \begin{bmatrix}-0.031\\-0.032\\-0.096\end{bmatrix} \tag{3.84}$$

Clearly, the error is much lower for the second point since this point is closer to the nodal point $[1,1]^T$. In order to demonstrate how the error of interpolation changes with the distance from nodes, let us define the error of interpolation as the length of the vector that is equal to the difference between the analytical solution and the interpolated one, namely in this particular case:

$$error = \left\| f(\mathbf{x}) - \begin{bmatrix}\sqrt{x_1}\\x_2^2\\x_1 + x_2\end{bmatrix} \right\| \tag{3.85}$$

Figure 3.6 visualizes the error of interpolation when linear spline RBF is used. It is visible that the error is "pinned" to zero for all the nodes and that it gradually increases with the increase of the distance from them. Clearly the potential maximum of the error of interpolation can be found in the points that are farthest from the nodes. However, depending on the nature of the function the error doesn't have to be the same in all the zones even if the distance from the nearest nodes is the same. In this case, the point with the coordinates $[2.5, 2.5]$ has somewhat larger error than the point $[1.5, 1.5]$ even though both are placed on equal distances from the nodes but in different zones of the domain. This behavior is also connected with

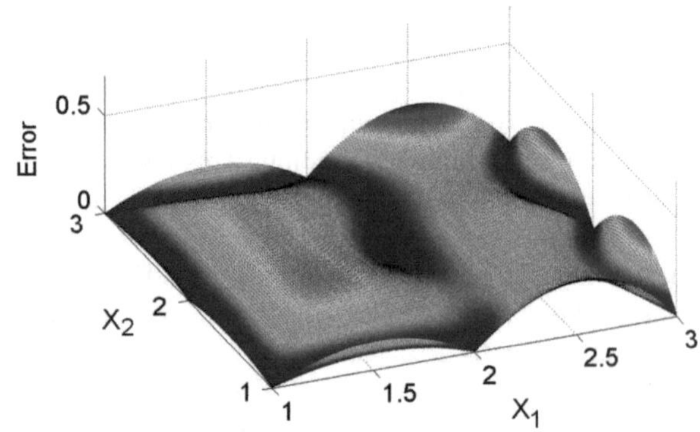

Fig. 3.6 Error of interpolation over the domain $x_1 = [1,3], x_2 = [1,3]$, with spline RBF

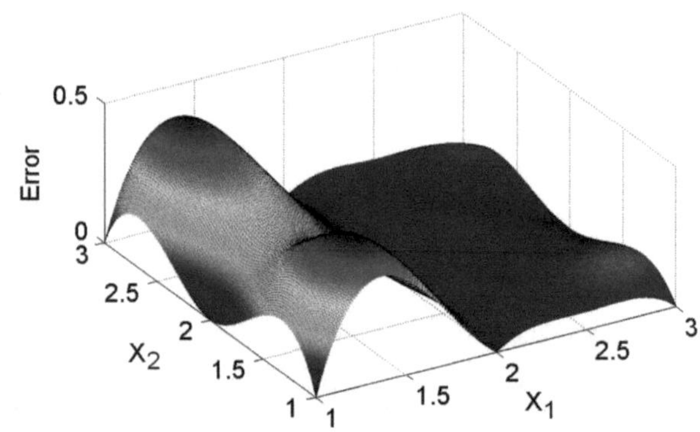

Fig. 3.7 Error of interpolation over the domain $x_1 = [1,3], x_2 = [1,3]$, with cubic spline RBF

the type of RBF used for the basis. The same error distribution for the previous example is computed for the cubic spline and it is visualized in Fig. 3.7.

The behavior of this type of RBF is quit opposite to the previous one, and shows much better results in the zone of $x_1 > 2$ than in the zone $x_1 < 2$.

Generally it can be concluded that the error of RBF interpolation certainly decreases by the increase of the number of nodes for which the value of the function is available. Once the number of nodes is chosen, the error can be further minimized by the right choice of the RBF type which should result from some comparative analyses.

3.6 POD-RBF Procedure

Up to know we have seen how on very effective way the set of previously generated responses of a certain system can be compressed using the discrete POD theory. This compression allows for a significant reduction of the dimensionality, preserving at the same time accuracy of the approximation. The approximation achieved in this way, refers only to a discrete number of system responses that are already computed using the "full" model (like for instance, FE model). A further step in order to make a continuous approximation of the system over certain parameter domain can be achieved if the POD is combined with the RBF interpolation. This brings a needed generalization to the low-order approximation, by building one function that approximates the response of the system.

To make the things more concrete, let us recall that our goal is to define an approximation that should be used instead of FE simulations. Therefore, we whish to find a function that depends on some parameters collected in vector **p** such that

$$f(\mathbf{p}) = \mathbf{u} \tag{3.86}$$

In the last equation vector **u** collects required output of the system. Since the function (3.86) should replace FE simulation, **u** can represent for example vector of nodal displacements, or any other value of interest resulting form FE simulations. This function is approximating the system response over some domain in parameter space. This practically means that for any arbitrary combination of parameters within this specified domain, in order to have approximation of system response, we can use function (3.86) instead of FE simulation.

The short description of RBF in the previous section suggested that this technique can be used for continuous approximation over some domain. In the view of applying RBF to the present purpose, let us recall that the snapshot matrix was generated using a certain number of input vectors $\mathbf{p}_i$, that here will represent nodes in the parameter space, for which the values of the function are known and are collected in the snapshot matrix **U**. However, since we already constructed a low-order approximation of these responses, they can be represented in the new, truncated system by the matrix of amplitudes $\bar{\mathbf{A}}$. This practically means that RBF can be applied in already reduced dimensionality where responses are expressed as the amplitudes, and so the function we are seeking for is the following one

$$f_a(\mathbf{p}) = \bar{\mathbf{a}} \tag{3.87}$$

Previously defined connection between the responses expressed in the reduced and full dimensionality which is given by (3.7), holds also for the values of functions f and f_a so we can write

$$f(\mathbf{p}) = \bar{\Phi} \cdot f_a(\mathbf{p}) = \mathbf{u} \tag{3.88}$$

Applying the RBF technique the approximation of f_a is written as a linear combination of some basis functions g_i, namely

$$f_a(\mathbf{p}) = \begin{bmatrix} b_{11} \\ b_{21} \\ \dots \\ b_{K1} \end{bmatrix} \cdot g_1(\mathbf{p}) + \begin{bmatrix} b_{12} \\ b_{22} \\ \dots \\ b_{K2} \end{bmatrix} \cdot g_2(\mathbf{p}) + \dots + \begin{bmatrix} b_{1N} \\ b_{2N} \\ \dots \\ b_{KN} \end{bmatrix} \cdot g_N(\mathbf{p}) = \begin{bmatrix} \bar{a}_1^i \\ \bar{a}_2^i \\ \dots \\ \bar{a}_K^i \end{bmatrix} \tag{3.89}$$

or written in matrix form

$$f_a(\mathbf{p}) = \begin{bmatrix} b_{11} & b_{12} & \dots & b_{1M} \\ b_{21} & b_{22} & \dots & b_{2M} \\ \dots & \dots & \dots & \dots \\ b_{K1} & b_{K2} & \dots & b_{KM} \end{bmatrix} \cdot \begin{bmatrix} g_1(\mathbf{p}) \\ g_2(\mathbf{p}) \\ \dots \\ g_M(\mathbf{p}) \end{bmatrix} = \mathbf{B} \cdot \mathbf{g}(\mathbf{p}) \tag{3.90}$$

Like in other examples of RBF interpolation, after the basis functions are chosen, we need to solve for the interpolation coefficients collected in matrix **B**. Having in mind that the values of the function f_a to be approximated are collected in the matrix of amplitudes **A** in the reduced space, Eq. 3.82 in this case has the form

$$\mathbf{B} \cdot \mathbf{G} = \bar{\mathbf{A}} \tag{3.91}$$

Last equation is solved for unknown matrix **B**, and then finally, combining (3.90) with (3.88) we arrive to the required general formula for the approximation of the system response for arbitrary parameter combination which reads

$$\mathbf{u} \approx \bar{\Phi} \cdot \mathbf{B} \cdot \mathbf{g}(\mathbf{p}) \tag{3.92}$$

To sum up, Eq. 3.92 represents an RBF approximation of the system response in the reduced space, which is transformed back to the original space (of full dimensionality). It was derived by performing the RBF interpolation of the system responses in the reduced space (represented by amplitude matrix $\bar{\mathbf{A}}$) and further transformed by pre-multiplying it by reduced POD basis $\bar{\Phi}$. Further in the book, this approximation will be called a POD-RBF approximation of the system response. The response computed like this has the original dimensionality, as the one computed by a "full" numerical model (i.e. FE model). It is obvious that (3.92) represents computationally "light" formulation since it involves the simple matrix multiplication. The accuracy of this approximation, as it will be demonstrated in the numerical examples that will follow, is practically on the level of FE simulations, but computed in a time shorter by several orders of magnitude.

For any arbitrary combination of parameters one only needs to compute vector **g** as a function of them, since matrices $\bar{\Phi}$ and **B**, are collecting constants computed once-for-all. The computation of these constants, hereafter called "training" of the model, involves a series of FE simulations but it is done only once. Figure 3.8 summarizes in the form of flow-chart all the operations needed to be performed within this process of training.

Previous example showed that, once trained, POD-RBF procedure can give results of practically the same accuracy as those computed by FEM. In this simple linear problem, only 16 analyses were enough to train POD-RBF procedure that gives an error of about 2%. Obviously adding additional analyses would reduce this difference, and could easily achieved values significantly smaller than 1%, as it will be demonstrated in further examples.

What should be emphasizes is that within POD-RBF procedure, the system response corresponding to any arbitrary values of parameters, is computed by a

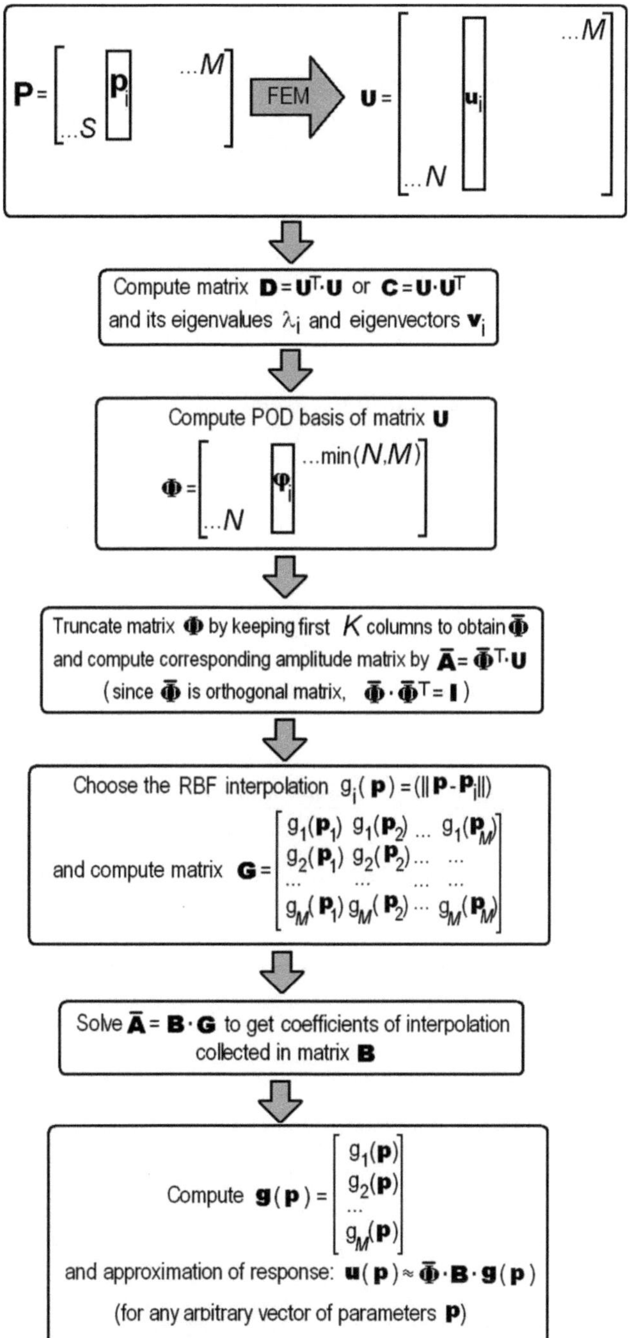

Fig. 3.8 Flow-chart of the operations needed to be performed in order to "train" POD-RBF model

Example 3.6. Consider the truss structure given in example 3.4. Train POR-RBF model that uses previously generated snapshot matrix for the nodal displacements and compare the results with those obtained by FE solution for another set of parameters (e.g. E = 135 GPa and A = 11 mm^2).

Solution:

To solve this problem apart of truncated POD basis and corresponding vector of amplitudes (here the basis was truncated to just one component) as additional information to compute the coefficients of interpolation collected in matrix **B** *we also need matrix* **P** *collecting parameter combinations used to generate snapshot matrix.*

Considering the range used in the example 3.4, we should first generate this matrix by the following loop in MATLAB

```
*********************************************************
cnt=0;
for E=125000:25000:200000
    for A=6:2:12
        cnt=cnt+1;
        P(:,cnt)=[E;A];
    end
end
*********************************************************
```

Note that this can also be done in the same loop in which the snapshot matrix is created in the listing given in Example 3.4

Using previously computed vector of amplitudes, we can compute matrix **B**. *For this purpose the following MATLAB routine should be written and executed*

```
*********************************************************
function [B]=podBmtx(A,p)
% Interpolation by the use of RBFs (Linear splines type)
% Normalization of P [0 1]
for j=1:size(p,1)
    minP(j,1)=min(p(j,:));
    maxP(j,1)=max(p(j,:));
    for i=1:size(p,2)
        x(j,i)=(p(j,i)-minP(j))/(maxP(j)-minP(j));
    end
end
N=size(x,2);
for i=1:N
    for j=1:N
        G(i,j)=sum((x(:,i)-x(:,j)).^2).^0.5;
    end
end
Bt=inv(G')*A';
B=Bt';
*********************************************************
```

When calling last routine as a function in MATLAB environment, as its arguments, vector of amplitudes **A** *and matrix of parameters* **P** *should be given, and the function will return matrix* **B**.

With this the training of procedure is finished, since the coefficients of matrices $\boldsymbol{\Phi}$ *and* **B** *are determined. For any further computation of system response, Eq.* 3.92 *can be used, where the only part that changes as a function of parameters is vector* **g** *which should be computed using the same type of RBF that was used to determine matrix* **B**. *This part is done within another routine listed below*

```
***********************************************************
function G=podGvec(p,pX)
% Function that constructs G vector as function of given
% parameters

N=size(p,2); % The number of generated snapshots
% Normalization of p
for j=1:size(p,1)
    minP(j,1)=min(p(j,:));
    maxP(j,1)=max(p(j,:));
    for i=1:size(p,2)
        x(j,i)=(p(j,i)-minP(j))/(maxP(j)-minP(j));
    end
end
gi=inline('(sum((x-y).^2).^0.5)');
value=pX;
for k=1:N
    G(k,1)=gi(value,x(:,k));
end
***********************************************************
```

Note that since the function works with normalized values of parameters (within the range of 0–1), also the parameters here given as `pX` *should be normalize in the same manner. Therefore, in this particular case, when the response needs to be computed for* $E = 135$ *GPa, and* $A = 11$ *mm*^{2}*, having in mind the range of training, vector of normalized parameters to be used as input to the function* `podGvec` *is computed by:*

$$\mathbf{pX} = \begin{bmatrix} \dfrac{135-125}{200-125} \\ \dfrac{11-6}{12-6} \end{bmatrix} = \begin{bmatrix} 0.133 \\ 0.833 \end{bmatrix}$$

The function gives as a result vector **g** *that can be finally used to compute the system response, which in this case is equal to:*

$$\mathbf{u} \approx \bar{\Phi} \cdot \mathbf{B} \cdot \mathbf{g}(\mathbf{p}) = \begin{bmatrix} -5.707 \\ -14.087 \\ -0.439 \\ -15.737 \\ 3.512 \\ -8.168 \\ -1.317 \\ -17.546 \\ -1.317 \\ -13.927 \end{bmatrix}$$

The response of the same system (i.e. for the same values of E and A) is computed by FEM using the listing given in the example 3.4. The resulting vector of nodal displacements is given below

$$\mathbf{u}_{FEM} = \begin{bmatrix} -5.836 \\ -14.407 \\ -0.449 \\ -16.094 \\ 3.591 \\ -8.353 \\ -1.347 \\ -17.944 \\ -1.347 \\ -14.244 \end{bmatrix}$$

Comparing the two results in may be observed that the maximum difference is equal to 0.39 mm, it occurs on the 8th degree of freedom and it represents approximately 2%.

The following figure visualizes the resulting displacements of the structure computed by POD-RBF procedure, and FEM. The displacements are magnified 10 times and the zoom is given for the node with the largest difference between the two methods.

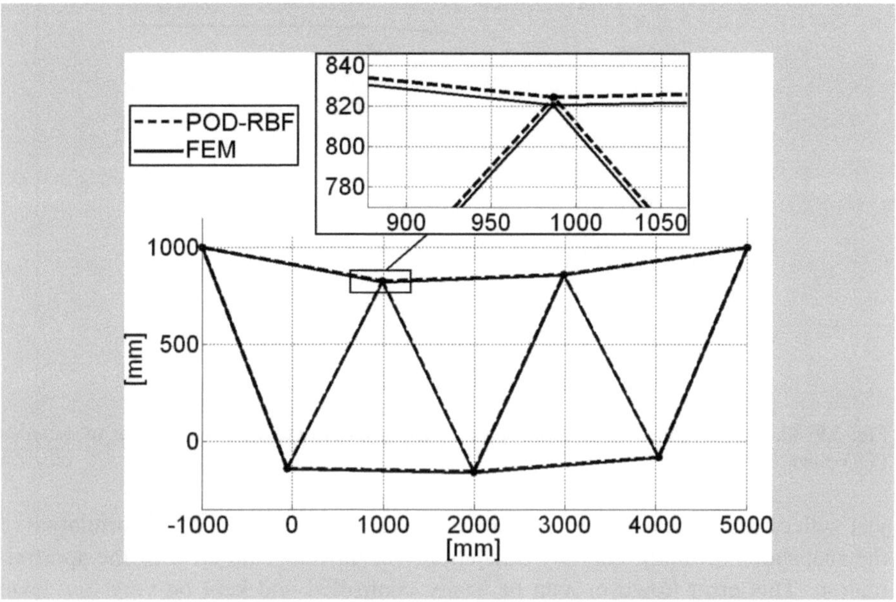

simple matrix multiplication. In previous, rather trivial example, there was no significant savings in the computing time. For more complicated models computing times by FEM can be extremely elevated, while the POD-RBF procedure, after the training, is always represented by a "light" computation, which can be just slightly longer due to the larger sizes of the matrices (i.e. for the systems with large number of degrees of freedom). Therefore the real advantage of this approach is visible on more complicated models and it grows with the complicity of the model.

3.7 On Sources of Error in Low-Dimensional POD-RBF Approximation

When we are talking about any low-dimensional approximations an important issue is also how accurate they are. In previous example we saw that the difference between full numerical model (FEM) and "light" one (POD-RBF) was about 2%. This section will summarize some general remarks on the estimation of error valid in general for POD-RBF approximation.

Considering the way it is built, POD-RBF procedure has two different sources of error. The approximation is based on a set of analyses previously performed, but the results of these analyses are not used directly in the low-order model. Taking advantage of the correlation of the responses (snapshots) the results are first transferred to another reference system, the POD basis, where their dimensionality is reduced by expressing them just with a few components. In the case of full correlation, as we already saw, they can be expressed without any loss of accuracy

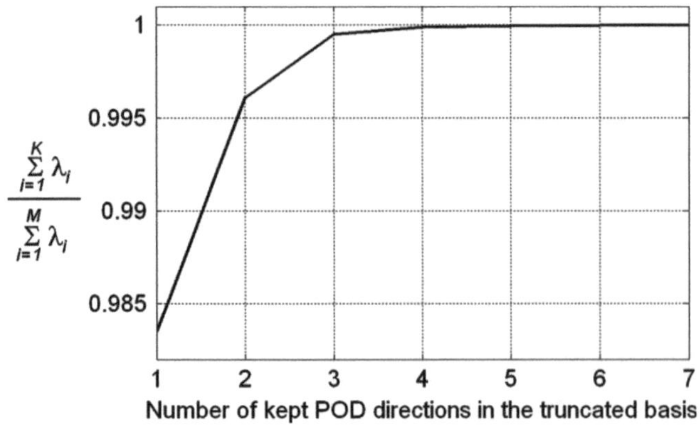

Fig. 3.9 Fast convergence of the summation of kept eigenvalues with the number of included POD directions

just with one component. In more realistic cases, there won't be a full correlation of the snapshots, meaning that any truncation will introduce an error of the approximation. This error however, can be easily controlled and kept on very low level using as criterion for truncation the ratio given by Eq. 3.31.

In practice, it is useful to construct the graph like the one visualized in Fig. 3.9, and to use it to determine the number of POD directions to be included in the reduced model. This graph gives the information about how the ratio changes when new directions are added. In the case visualized here, there is a strong correlation between the snapshots, since the summation practically reaches value 1 when just five directions are included, but there is no significant difference even when the low-dimensional model with three directions is used. In the case when snapshot matrix collects the velocities, it can be shown that (see [8] and [23]) eigenvalue λ_i is equal to the kinetic energy transferred by the ith POD mode. In this case there is also a physical meaning of the graph given in Fig. 3.9, since then, the ordinate is showing what is the amount of kept kinetic energy of the low-order model with respect to the full one. Consequently, the summation of neglected eigenvalues corresponds to the loss of energy introduced by the approximation. This is the reason why sometimes in the literature the eigenvalues are also called *energies* of the POD modes. For other cases there is no physical meaning of the eigenvalues, but in any case the ratio (3.31) provides quantitative information about the accuracy of the approximation, since in the case when it equals one, there is no error and the snapshots are expressed without any loss of information. Therefore, to control this part of the error one should use the number of directions for which the ratio (3.31) is almost 1 (e.g. equal to 0.999). It means that this part of error can actually be totally eliminated since one can always choose the number of directions large enough for the reconstruction of the snapshots without practically any loss.

Up to this stage, the approximation can be achieved only for the set of parameters for which the results of the full model are previously computed. In order to have a formulation that can be compared to the FE simulations, it is required to attribute a so-called generalization feature to the approximation. This practically means that it is required to have a procedure capable of computing system responses also for those parameter combinations that were not used in the phase or training. This is achieved, as we already saw, by implementing an RBF interpolation of the data in the space of reduced dimensionality.

Applying this technique, we will have one function (given by Eq. 3.92) that approximates the system response. From this stage, no further references will be made to the original snapshot matrix, or its amplitude matrix corresponding to the reduced space. The system response can now be calculated for any parameter combination, including those previously used in the training by Eq. 3.92. However, having in mind that interpolation coefficients within RBF technique are determined such that the approximation of the function is exactly satisfied in the "nodes" (here parameter combination used in the training) the approximation (3.92) can therefore reconstruct the responses exactly in the training points. This means that the further approximation introduced by RBF is not altering the previously established level of error in these points, and it still will be controlled by the ratio (3.31). Therefore the second source of error, due to the RBF interpolation will be visible only in those points that are not coinciding with the training points. Further, this error will be directly proportional to the distance, in the parameter space, between the point for which the response is approximated and the nearest training points (like it is visualized in Figs. 3.6 and 3.7). It implies that the error will be minimized by keeping the distance between the training points small.

In practical problems, the easiest way to control an interpolation error is to use, where possible, the regular grid of nodes in parameter space to perform the training. Nevertheless, a good feature of this approximation is that anyhow the zone of maximum error can be evidenced. Of course this error can be always further reduced by enriching the set of analyses used for the training. However the choice of the total number of analyses required to be performed in order to calibrate POD-RBF procedure of acceptable accuracy is connected with the problem itself, and there are no specific guidelines for its selection. Nevertheless following examples will demonstrate that with the reasonable number of training analyses, also in non-linear problems, good results can be achieved with this procedure.

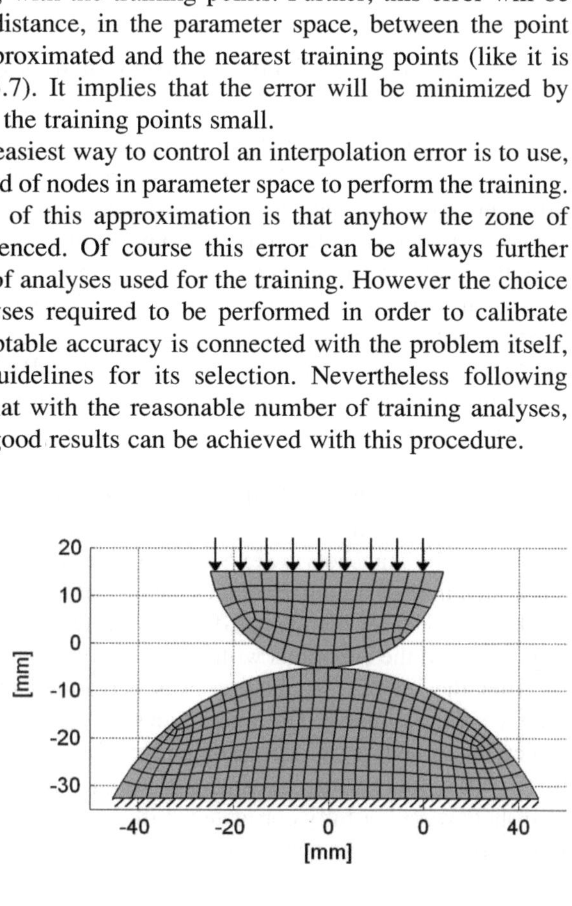

Fig. 3.10 FE model of two cylinders in radial contact

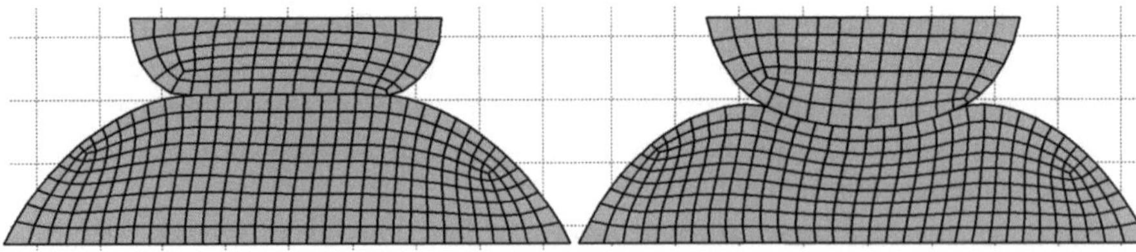

Fig. 3.11 Different resulting displacements depending on the stiffness of two cylinders

3.8 Examples of the Use of POD-RBF Procedure for Fast Simulation

This section will present some numerical examples in which previously presented technique will be trained and used for different problems. The results will be compared to those obtained by traditional FEM simulations. In all of the examples, POD-RBF procedures were trained using the commercial code ABAQUS for FE simulations.

3.8.1 Example 1: Two Cylinders in Radial Contact

In this example the POD-RBF procedure is trained and used to compute nodal displacements of 2D model of two cylinders in radial contact. Cylinders are made of different materials, both of them assumed to deform only in the elastic range. Bigger cylinder is fixed at the bottom, while the smaller one is pushed against it by the load applied on its top surface, as indicated in Fig. 3.10.

The snapshot matrix collects the nodal displacement of both cylinders, and is constructed by varying both Young's moduli (E_1 of the smaller cylinder and E_2 of the bigger one) in the range from 50 to 200 GPa with the step of 10 GPa. For the training points a regular grid was used, with the total number of them being equal to 256. The simulations were performed using a FE model with 818 degrees of freedom.

Even though to both cylinders it was attributed linear elastic behavior, due to the presence of the unilateral contact, the response of the structure is non-linear. Moreover, the snapshots are expected to be less correlated than in previous example since the range of training includes passing of the structure from the situation when the bigger cylinder is much stiffer (having the response like the one presented in Fig. 3.11a) to the situations with much stiffer smaller cylinder (Fig. 3.11b).

After the set of 256 analyses was performed the POD basis of the snapshot matrix was computed. The magnitudes of the eigenvalues associated with POD directions turned out to decrease rapidly. The values of the first five of them were: $9.391 \cdot 10^7$, $1.441 \cdot 10^4$, 15.169, 0.215 and 0.066. Even though as expected the snapshots were not fully correlated, still the fast drop of the magnitudes confirmed that there was a strong correlation of them. The basis was truncated keeping just the

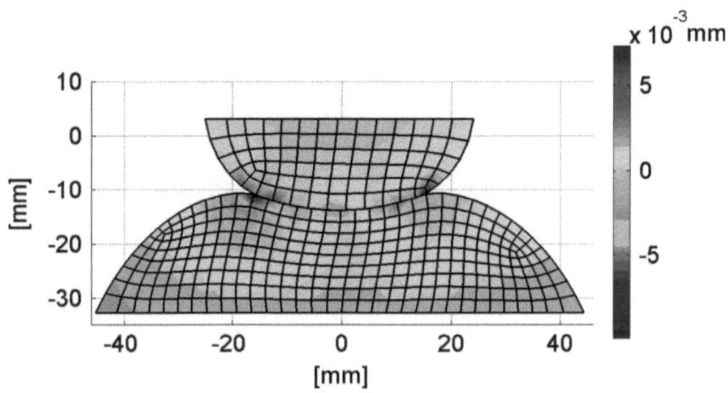

Fig. 3.12 Difference between nodal displacements computed by reduced model (POD-RBF) and full model (FEM)

first three components, and the ratio between sum of all neglected eigenvalues and the sum of all of them was smaller than 10^{-8}, a quantity that is giving information about the accuracy of low-dimensional approximation of originally computed snapshots. All 256 snapshots previously computed by a full FE model, can be expressed practically without any loss. At the same time the dimensionality is reduced from original 818 (that equals number of degrees of freedom of used FE model) to just three components.

Further interpolation of thus prepared data was performed by adopting a cubic-spline RBF defined by Eq. 3.74. Therefore, the training process resulted in two matrices, a [818×3] POD basis matrix $\bar{\Phi}$ and a [3×256] matrix $\mathbf{B}$. These matrices are collecting constants, while the vector $\mathbf{g}$ is a function of parameters, and is computed each time when the response for arbitrary values of parameters is required. Their multiplication according to (3.92) results in nodal displacements of a full structure consisting of two cylinders.

In order to check the overall error of the approximation nodal displacements are computed for the pair of parameters that was not considered in the training phase. Parameter values are chosen to be equal to $E_1 = 185$ GPa and $E_2 = 65$ GPa. This choice intended to maximize the part of the error due to RBF interpolation by selecting the farthest parameters from grid nodes used in the training (recall that a regular grid with a step 10 GPa was used).

The same problem is solved by FE method in order to compare the results obtained by full and reduced model. Figure 3.12 visualizes the difference in nodal displacements between the two models

As it can be seen from the figure, in most parts of the model there is practically no difference at all in the displacements as most of the structure is in yellow-colored zone. The largest difference occurs in the zones where the absolute displacement is of the order of couple of millimeters, and even there it is three orders of magnitudes smaller, being equal to about 5 μm, that corresponds to about 0.1%. This practically confirmed that the error of approximation was negligible and that POD-RBF

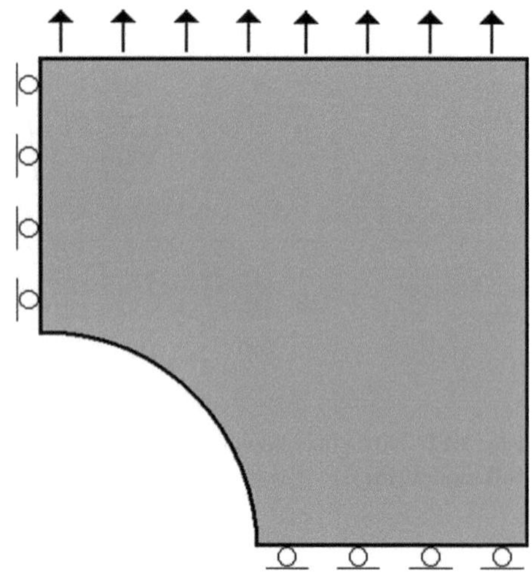

Fig. 3.13 One-quarter of plate with circular hole

procedure could produce the same results as FE model. Comparing the computing times in this particular case since FE model was not complicated the difference was about 1,000 times. Already here it is evidenced that the efficiency of low-dimensional POD-RBF procedure becomes more noticeable by growing complexity of the model.

3.8.2 Example 2: Plate with Circular Whole

It was already mentioned that the POD-RBF procedure can be used to approximate any physical field. The previous example showed the approximation of displacement field, but also other values can be considered. Since the training process requires a set of simulations, after they are performed practically any field of possible interest can be approximated by the POD-RBF procedure with the same numerical "cost". The following example will show the approximation of the stress field together with the displacement-force curve of one particular structure.

Consider the structure shown in Fig. 3.13 that represents one quarter of rectangular plate with the circular hole in the center. The behavior of material is assumed to be elastic-perfectly plastic according to von Mises criterion. In this example two different POD-RBF approximations were constructed. The first one computes von Mises stress distribution in the Gauss points, while the second provides the reaction force at constrained edge versus the displacement of moving edge. The parameters that were changed across the training process are: Young's modulus E within the range of 100–200 GPa with the step of 10 GPa, and the yield limit σ_y in the range between 250 and 450 MPa, with the step of 10 MPa. A regular grid of training points in the parameter space is used, having the total number of them equal to 231.

In every analysis the same FE model was used, with exactly the same boundary conditions.

The first constructed snapshot matrix $\mathbf{U}_1$ was the one containing the stresses. Since the model was discretized with 801 four-node quadrilateral elements with four Gauss points used for numerical integration, the size of this matrix was $3{,}204 \times 231$. The first five eigenvalues of corresponding matrix $\mathbf{D}_1$ have the following magnitudes: $2.118 \cdot 10^{11}$, $1.215 \cdot 10^{7}$, $1.0571 \cdot 10^{6}$, $4.415 \cdot 10^{4}$ and $1.137 \cdot 10^{4}$. Truncated model consisted of just first three directions. The ratio between sum of neglected eigenvalues and sum of all of them was less than 10^{-5}.

The second POD-RBF approximation was trained to give as an output the relationship of reaction force versus imposed displacement. For that purpose, a snapshot matrix $\mathbf{U}_2$ was constructed collecting as columns 50 pairs of coordinates (displacements and forces) used to defined displacement vs. force curve. This matrix was therefore of the size 100×231. The magnitudes of first six eigenvalues of the corresponding matrix $\mathbf{D}_2$ turned out to have the following values: $7.338 \cdot 10^{12}$, $1.242 \cdot 10^{10}$, $4.627 \cdot 10^{8}$, $4.767 \cdot 10^{7}$, $9.218 \cdot 10^{6}$ and $2.415 \cdot 10^{6}$. Truncating the basis to the first five directions the same level of error as for $\mathbf{U}_1$ matrix was achieved.

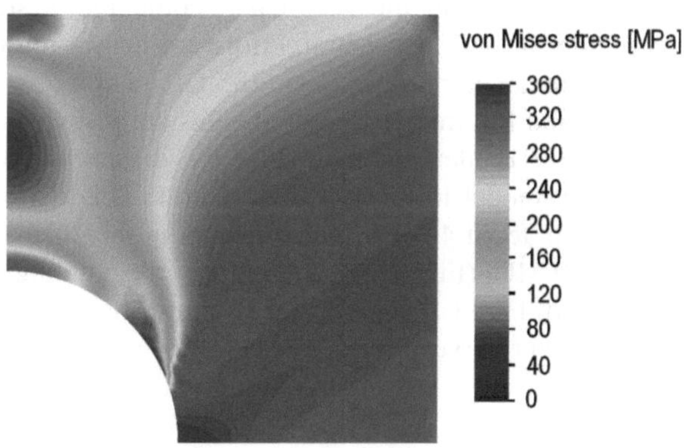

Fig. 3.14 Distribution of equivalent von Mises stresses computed by reduced POD-RBF model

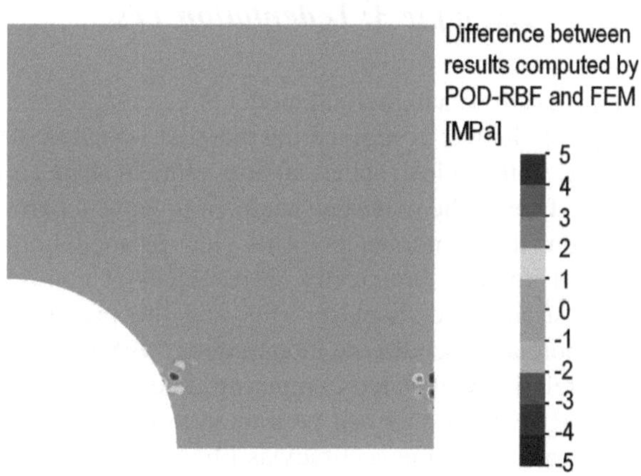

Fig. 3.15 Difference between POD-RBF and FEM results

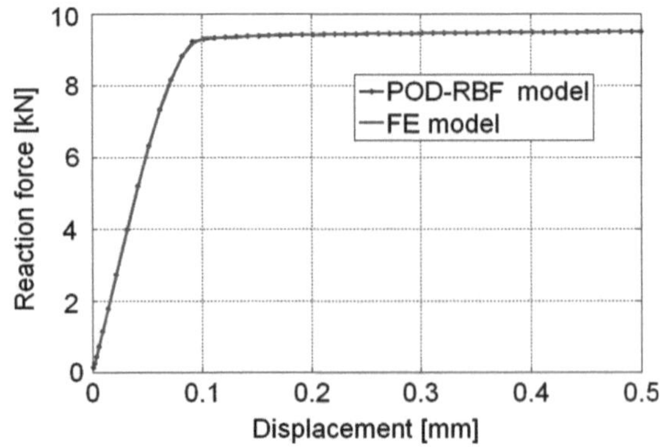

Fig. 3.16 Force Vs displacement curve computed by POD-RBF and FEM

Also in this example a cubic-spline was used to perform RBF interpolation. After the interpolation constants are determined the whole training process was finished resulting in four matrices with constants: two for the approximation of stress distributions $\bar{\Phi}_1$ [3,204×3] and $\mathbf{B}_1$ [3×231], and two for the force vs. displacement curve $\bar{\Phi}_2$ [100×5] and $\mathbf{B}_2$ [5×231]. For any arbitrary combination of parameters, after the vector $\mathbf{g}$ is computed, system response is approximated according to (3.92) by a straightforward matrix multiplication. The response considering parameter values, $E = 185$ GPa and $\sigma_Y = 310$ MPa was computed with reduced and full model, and the results are visualized in the Figs. 3.14–3.16.

Figure 3.15 demonstrated that the difference between "full" and "reduced" numerical model in most of the structure is equal to zero. The largest difference is equal to about 4 MPa, which occurs in the zones with absolute values of stress about 300 MPa (Fig. 3.14), therefore corresponding to a bit more than 1%. On the other hand, Fig. 3.16 shows that both models produced practically the same force-displacement curves.

3.8.3 *Example 3: Indentation Test*

In this example a numerical model of indentation test is considered. This, nowadays very popular test in engineering practice, is a process in which the tip of an indenter, usually with conical, spherical or pyramidal shape, is quasi-statically forced against the surface of the material specimen to leave a permanent imprint.

From the numerical point of view its modeling represents a challenging task since it involves practically all resources of non-linearity (material non-linearity, large strains and displacements and the presence of contact). Using available commercial FE codes the modeling of this test can be done fairly well, but usually with relatively elevated computing times. In this section we will show that a fast POD-RBF procedure can be trained to predict the response from such a test with practically the same accuracy as FEM.

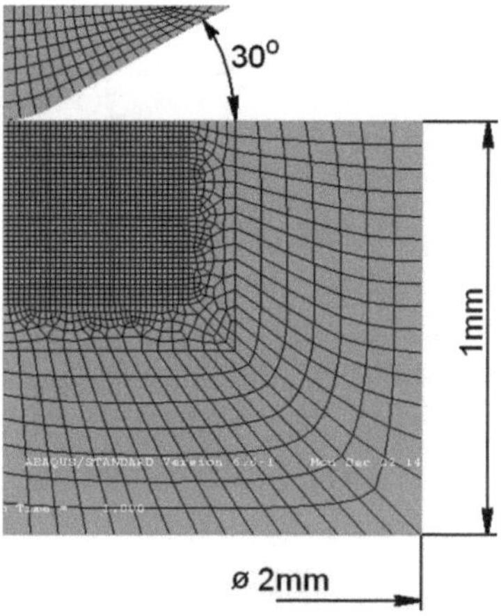

Fig. 3.17 FE model of conical indentation test

Indentation test with standard Rockwell conical indenter was modeled. The indenter is considered deformable with the following elastic properties: Young modulus of 1,170 GPa and Poisson ratio equal to 0.07. FE model used to perform the training consisted of 1,715 four-node quadrilateral elements (Fig. 3.17). Material model attributed to the indented specimen was elastic, perfectly plastic with von Mises yield criterion. While Poisson ratio is a priori assumed, $\nu = 0.3$, the Young modulus E and the yield stress σ_Y are varied in the phase of training within the following ranges: from 140 to 220 GPa and from 300 to 500 MPa with the steps of 10 GPa and of 10 MPa respectively. The training therefore resulted in the total of 189 analyses.

The fields of interest that are approximated by POD-RBF model are the components of the stresses σ_{11}, σ_{22}, equivalent von Mises stress and nodal displacements. Snapshots corresponding to these fields are collected in the following four matrices: 6,860×189 $\mathbf{U}_{11}$, 6,860×189 $\mathbf{U}_{22}$, 6,860×189 $\mathbf{U}_{vm}$ and 3,556×189 $\mathbf{U}_{D}$.

For the truncation the criterion was adopted of reducing to less than 10^{-6} the ratio between summation of neglected eigenvalues and summation of all of them. This criterion led to a different number of POD directions preserved for each one of the four above specified snapshot matrices. The most correlated data turn out to be those collecting nodal displacements. In this case only the first four new reference axes were sufficient to satisfy the above specified accuracy criterion. For stress fields, this number was slightly larger and the bases are truncated after the 14th, 10th and 12th directions, for von Mises stress σ_M,' for stress component σ_{11} and stress component σ_{22} respectively. Considering that the original dimensionality was 3,556 for the displacement field, and 6,860 for the stress fields, it is obvious that in all of the cases the reduction was significant.

As for the RBF interpolation, here the inverse multiquadric function was adopted given by the following formula

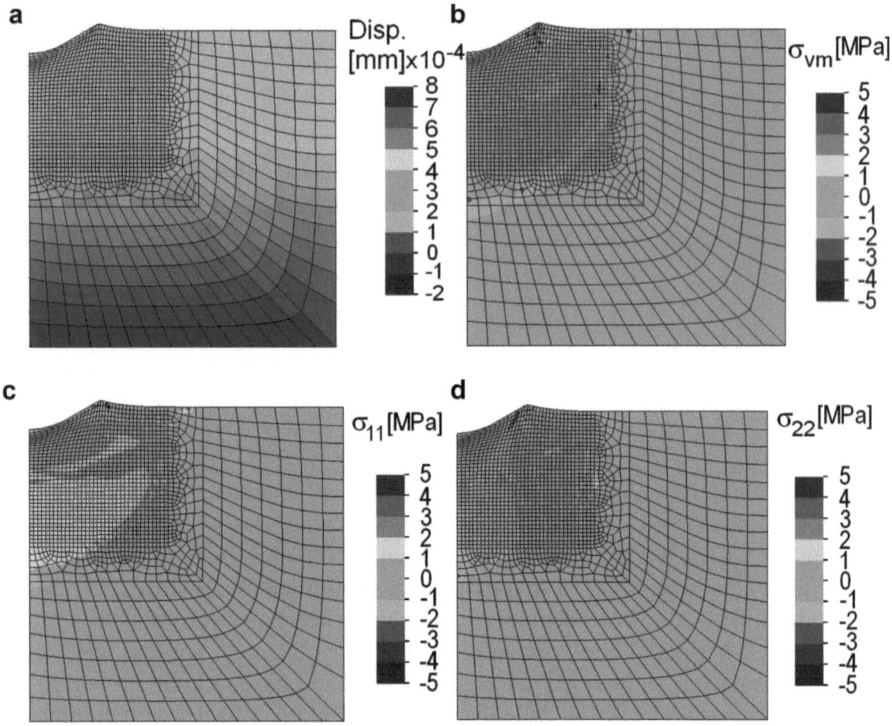

Fig. 3.18 Differences between displacement fields and distribution of stresses computed by FEM and POD-RBF

$$g_i(\mathbf{p}) = \frac{1}{\sqrt{\|\mathbf{p} - \mathbf{p}_i\|^2 + r^2}} \tag{3.93}$$

where r is a "smoothing coefficient", which has been assumed herein equal to 0.5. Note that also in this example parameters (i.e. entries of vector $\mathbf{p}$) are normalized to be in the range 0–1 (where 0 corresponds to the lower bound and 1 corresponds to the upper bound).

The verification of the accuracy of reduced model is performed on the couple of parameters not considered in the training, specifically $E = 195$ GPa and $\sigma_Y = 405$ MPa. For these parameters the test was simulated by both FEM and POD-RBF model. Figure 3.18 visualizes comparative results.

From the figure it may be observed that the differences in the distribution of the stresses just in some small zones have the value of about 4 MPa. Considering that these zones are in the region where the material yields (for this example yield limit was 405 MPa) it is obvious that this difference is less than 1%. As for the displacements (Fig. 3.18a), the maximum difference is less than 1 μm.

For this model the reduction of computing time was huge, since the FEM was already complicated enough. In this particular case POD-RBF procedure was about 30,000 times faster. This practically means that, once that the long phase of training is finished, the results of the simulation of indentation test can be computed practically in real-time.

3.9 Summary

In this chapter it was shown how to construct a POD approximation of a discrete data. Three different methods of building a POD basis for the given data set were presented evidencing that they are leading to exactly the same results. It was shown how to use this mathematical tool for the approximation of discrete fields, like those computed by numerical models in structural analyses.

Since the goal here was to develop a general formulation that could be used as a replacement for the standard numerical models (like FEM), an RBF interpolation was performed on data with previously reduced dimensionality by POD. This non-local type of interpolation provides a single matrix equation that is a function of some parameters for which it is “trained”. It was shown how to derive a general formulation that combines these two mathematical procedures to be used for a prediction of system responses. This POD-RBF model was referred to as low-dimensional or “reduced” model and it is represented by a simple matrix multiplication where one of the vectors involved in it is a function of parameters. This practically means that in order to predict the system response for any arbitrary values of parameters, instead of using “full” and costly numerical model (like FEM) one can use this, computationally light formulation. The exercises presented in the last part of the chapter showed that “reduced” model is capable to produce almost the same results as the “full” one.

All of the examples showed a strong correlation of the snapshots which allowed for significant reduction of the dimensionality without practically any loss of accuracy. This can be explained by the fact that the snapshots represent the outputs of the same system when just some of the parameters are changed. For instance, the last example showed the prediction of the indentation test of the same specimen size, same indenter geometry, same load and same boundary conditions corresponding to different materials (same material model but different constants). This similarity between the simulations expressed itself in a strong correlation between the results and so the POD approximation could be effectively applied.

It should be emphasizes however that in order to calibrate the reduced model a set of costly analyses needs to be performed. Some of the examples treated in this chapter involved couple of hundreds of analyses. Therefore, it is quite obvious that it is not computationally justifiable to “train” the POD-RBF model and to use it afterwards for a single computation.

On the other hand, parametric studies and identification problems based on inverse analyses require performing repeated numerical analyses, where only a few parameters are varied among those that are uniquely defining the problem. This class of problems can gain significant advantages of the use of such a fast computational tool like the one presented in this chapter.

References

1. Pearson, K.: On lines planes of closes fit to system of points in space. The London, Edinburgh Dublin Philos. Mag. J. Sci. **2**, 559–572 (1901)
2. Hotelling, H.: Analyses of complex statistical variables intro principal components. J. Educ. Psychol. **24**, 417–441 (1933)
3. Karhunen, K.: Uber linear Methoden fur Wahrscheiniogkeitsrechnung. Ann. Acad. Sci. Fennicae Series Al Math. Phys. **37**, 3–79 (1946)
4. Loeve, M.M.: Probabilty Theoiry. Van Nostrand, Princeton (1955)
5. Lumley, J.L.: Stochastic Tools in Turbulence. Academic, New York (1970)
6. Liang, Y.C., Lee, H.P., Lim, S.P., Lin, W.Z., Lee, K.H., Wu, C.G.: Proper orthogonal decomposition and its applications: part I – theory. J. Sound Vib. **252**(3), 527–544 (2002)
7. Bialecki, R.A., Kassab, A.J., Fic, A.: Proper orthogonal decomposition and modal analysis for acceleration of transient FEM thermal analysis. Int. J. Numer. Meth. Eng. **62**, 774–797 (2005)
8. Holmes, P., Lumley, J.L., Berkoz, G.: The proper orthogonal decomposition in the analysis of turbulent flows. Annu. Rev. Fluid Mech. **25**, 539–575 (1993)
9. Kerschen, G., Ponceletm, F., Golinval, J.C.: Physical interpretation of independent component analysis in structural dynamics. Mech. Syst. Signal Process **21**, 1561–1575 (2007)
10. Ly, H.V., Tran, H.T.: Modeling and control of physical processes using proper orthogonal decomposition. Math. Comput. Model **33**, 223–236 (2001)
11. Ruotolo, R., Surace, C.: Using SVD to detect damage in structures with different operational conditions. J. Sound Vib. **226**(3), 425–439 (1999)
12. Sirovich, L., Kirby, M.: Low-dimensional procedure for the characterization of human faces. J. Opt. Soc. Am. **4**, 519–524 (1987)
13. Tang, D., Kholodar, D., Juang, J.N., Dowell, E.H.: System identification and proper orthogonal decomposition method applied to unsteady aerodynamics. AIAA J. **39**(8), 1569–1575 (2001)
14. Jolliffe, I.T.: Principal Component Analysis. Springer, New York (2002)
15. Golub, G.H., Van Loan, C.F.: Matrix Computations. The Johns Hopkins University Press, Baltimore/London (1993)
16. Ostrowski, Z., Bialecki, R.A., Kassab, A.J.: Solving inverse heat conduction problems using trained POD-RBF network. Inverse Probl. Sci. Eng. **16**(1), 705–714 (2008)
17. Buljak, V.: Assessment of material mechanical properties and residual stresses by indentation simulation and proper orthogonal decomposition. Ph.D. thesis, Politecnico di Milano, Milano (2009)
18. Buljak V., Maier G.: Proper orthogonal decomposition and radial basis functions in material characterization based on instrumented indentation. J. Eng. Struct. (2009, submitted)
19. Bolzon, G., Buljak, V.: An indentation-based technique to determine in-depth residual stress profiles by surface treatment of metal components. Fatigue Fract. Eng. Mater. Struct. (2010, in press)
20. Buhmann, M.D.: Radial Basis Functions. Cambridge University Press, Cambridge (2003)
21. Aoki, S., Amaya, K., Sahashi, M., Nakamura, T.: Identification of Gurson's material constants by using Kalman filter. Comput. Mech. **19**, 501–506 (2007)

22. Kansa, E.J.: Motivations for using radial basis functions to solve PDEs. http://rbf-pde.uah.edu/kansaweb.pdf, pp. 1–8 (2001)
23. Holmes, P., Lumley, J.L., Berkoz, D.: Turbulence, coherent structures, dynamical systems and symmetry. Cambridge Monographs on Mechanics. Cambridge University Press, Cambridge, UK (1996)

Chapter 4
Inverse Analyses in Structural Problems: Putting All the Pieces Together

Inverse analyses procedures in structural problems are usually designed in order to assess some of the unknown parameters. Up to now we already saw that, for a successful development of fully operative inverse analysis procedure, it is required to put together three different elements: experimental technique, numerical simulation of it, and an optimization algorithm. The most traditional approach to the inverse analyses procedures, when structural problems are in focus, assumes that the simulation of the experiment is done by finite element modeling. It is classical, and the most common way of proceeding since nowadays there are well developed FE techniques at our disposal which can be used to simulate even complicated phenomena that may take place in a selected experiment. Given the required repeatability of the simulations enforced by the adopted optimization algorithm, this approach may not be the most convenient for the routine use, as it can be time consuming. Therefore, a modern approach to inverse analyses goes in the direction of avoiding a need to perform FE simulations every time when the inverse problem needs to be solved. One of the possible solutions of this problem is based on POD-RBF algorithm described in Chap. 3. Its implementation within an inverse analyses procedure will be discussed in the subsequent chapter. This chapter will focus on a traditional approach relaying on FE simulations, as it is anyhow very important at least in some of the phases of procedure development.

Numerically speaking, in order to develop an inverse analyses procedure, one needs to write a code to solve minimization problem, to develop a numerical model to simulate the test, and to make these two communicating. As for the first part, codes developed in Chap. 2 can be used rather successfully for the structural problems here of interest.

Within traditional approach as mentioned above, the numerical simulations are performed by the use of FE method, and in most of the cases it is convenient to use a commercial code. In order to use optimization algorithms developed in Chap. 2, it is required also to code a so-called objective function in a least squares form that will quantify the discrepancy between experimental and numerical data. As we already saw in the simple problems tackled in Chap. 2, these codes need to load data form the experiment, to compute their numerical counter-part, and to construct vector of

V. Buljak, *Inverse Analyses with Model Reduction*, Computational Fluid and Solid Mechanics, DOI 10.1007/978-3-642-22703-5_4,

residuals. Therefore for the structural problems here of interest, when the experiment is simulated by a commercial FEM software, the developed objective function code needs to be able to modify FE model by changing the parameters (that are arguments of the objective function), to run a FE simulation, and afterwards to load the results and to compare them with experimental ones in order to form vector of residuals.

For the case studies that will be presented in what follows a commercial code ABAQUS [1] is selected to perform FE simulations. This code has a possibility to define completely numerical model for the analyses by a human readable ASCII file. This feature is rather useful for the present purpose as here it is needed to perform series of different analyses in which just some of the parameters need to be modified (e.g. parameters entering into the constitutive model). It is therefore easy to make a MATLAB code (or code in any other programming language) that will load this text file, change what needs to be changed and save it for the further analyses. Therefore, for the following case studies, a listing of MATALB code will be given that solves the inverse problem completely automatically, without any intervention of the user once the FE model is built.

4.1 Case Study: Assessment of Two Elastic Parameters for the Sandwich Cantilever

Let us imagine that we need to build an inverse analysis procedure in order to assess Young's moduli of two different materials forming the sandwich cantilever like the one presented in Fig. 4.1. Let us further imagine that it needs to be done in the form already embedded together in a single structure like the one visualized in the figure.

The first step, as explained in Chap. 1, is to establish the experiment that we would like to use. For this very simple problem it is possible to think of a simple experiment in which en external edge of the cantilever will be loaded with known force, and the resulting displacements, say at the upper edge of the structure, can be

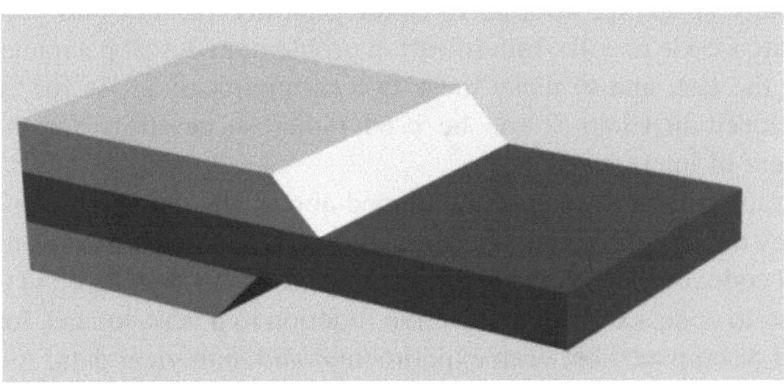

Fig. 4.1 Sandwich cantilever built of two different materials

measured. The problem can be expressed as two dimensional, and schematic representation of the adopted experiment is given in Fig. 4.2.

For this simple example it is quit intuitive to see that the proposed measurable quantities are influenced by the parameters and so there is no need to perform sensitivity analyses in order to verify that the experimental setup is adequate for the required task.

Adopting this experiment as the resource of information that will be further exploited by the inverse analyses procedure, a second step is to construct a numerical model of it. As mentioned above, within this case study a commercial code ABAQUS will be used to perform simulations of the experiment.

4.1.1 FE Model of the Experiment

The problem can be considered as 2D plane stress. The adopted mesh is visualized in Fig. 4.3. For this simple case only elastic material behavior is assumed, as the parameters of interest are the elastic constants, and therefore, the experiment should introduce loads that will keep the structural response within the elastic range.

In order to establish easier connection between MATLAB surrounding and ABAUQS we will work only on input file. Furthermore, to make our job easier, input file for the FE analysis will be divided in couple of parts, so that the information considering material constitutive model will be in a separate files. In such way MATLAB procedure that we will further design will load and change only these files.

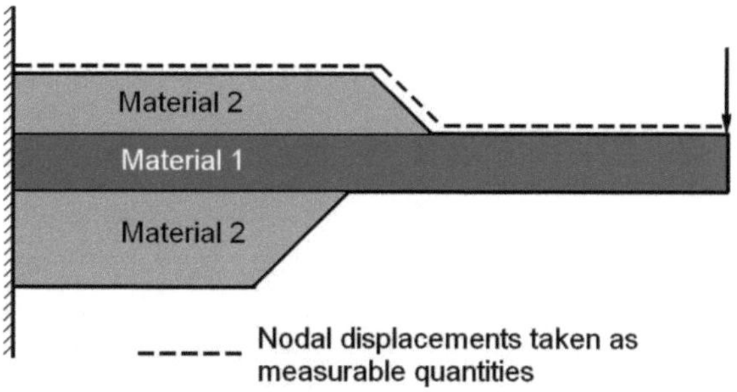

Fig. 4.2 Schematic representation of the adopted experiment

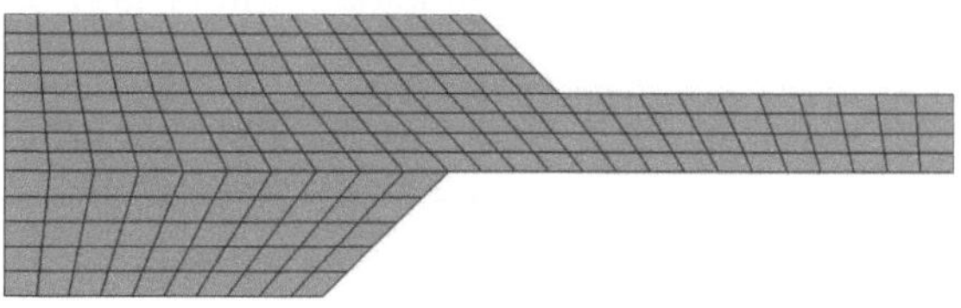

Fig. 4.3 FE mesh for the sandwich cantilever experiment

The main input file to the analyses has the following form:

```
*Heading
** Job name: cantl_cae Model name: Model-1
*Preprint, echo=NO, model=NO, history=NO, contact=NO
**
** PARTS
**
*Part, name=cantilever
*End Part
**
**
** ASSEMBLY
**
*Assembly, name=Assembly
**
*Instance, name=cantilever-1, part=cantilever
*Node
*Include, input=nodes.inp
*Element, type=CPS4
*Include, input=elements.inp
*Nset, nset=_PickedSet4, internal
   1,    2,    4,    6,    7,    8,    9,   10,   11,   12,   13,   14,
15,  16,  17,  18, 19,  37,  38,  39,  40,  41,  42,  43,  44,
45,  46,  47,  48,  49,  50,  63, 64,  65,  66,  67,  68,  69,
70,  71,  72,  73,  74,  75,  76,  77,  78,  79, 80,  81,  82,
83,  84,  85,  86,  87,  88,  89,  90,  91,  92,  93,  94,
95, 96,  97,  98,  99, 172, 173, 174, 175, 176, 177, 178, 179,
180, 181, 182, 183, 184, 185, 186, 187, 188, 189, 190, 191,
192, 193, 194, 195, 196, 197, 198, 199, 200, 201, 202, 203,
204, 205, 206, 207, 208, 209, 210, 211, 212, 213, 214, 215,
216, 217, 218, 219, 220, 221, 222, 223, 224, 225, 226, 227,
228, 229, 230, 231, 232, 233, 234, 235, 236, 237, 238, 239,
240, 241, 242, 243, 244, 245, 246, 247, 248, 249
*Elset, elset=_PickedSet4, internal, generate
 101,  210,    1
*Nset, nset=_PickedSet5, internal
   1,    2,    3,    4,    5,    6,   11,   12,   13,   14,   15,   16,
17,  18,  19,  20, 21,  22,  23,  24,  25,  26,  27,  28,  29,
30,  31,  32,  33,  34,  35,  36, 37,  38,  39,  40,  41,  42,
43,  44,  45,  46,  47,  48,  49,  50,  51,  52,53,  54,  55,
56,  57,  58,  59,  60,  61,  62, 100, 101, 102, 103, 104, 105,
106, 107, 108, 109, 110, 111, 112, 113, 114, 115, 116, 117,
118, 119, 120, 121, 122, 123, 124, 125, 126, 127, 128, 129,
130, 131, 132, 133, 134, 135, 136, 137, 138, 139, 140, 141,
142, 143, 144, 145, 146, 147, 148, 149, 150, 151, 152, 153,
154, 155, 156, 157, 158, 159, 160, 161, 162, 163, 164, 165,
166, 167, 168, 169, 170, 171
*Elset, elset=_PickedSet5, internal, generate
  1, 100,    1
** Region: (mat1:Picked)
*Elset, elset=_PickedSet5, internal, generate
  1, 100,    1
```

```
** Section: mat1
*Solid Section, elset=_PickedSet5, material=material1
1.,
** Region: (mat2:Picked)
*Elset, elset=_PickedSet4, internal, generate
 101,  210,    1
** Section: mat2
*Solid Section, elset=_PickedSet4, material=material2
1.,
*End Instance
**
*Nset, nset=_PickedSet37, internal, instance=cantilever-1
  1,  6,  7, 10, 60, 61, 62, 63, 64, 65, 66, 97, 98, 99
*Elset, elset=_PickedSet37, internal, instance=cantilever-1
   1,  26,  51,  76, 110, 120, 130, 140, 150, 151, 166, 181,
196
*Nset, nset=_PickedSet38, internal, instance=cantilever-1
 3,
*Nset, nset=upper, instance=cantilever-1
 4,  5,  9, 10, 51, 52, 53, 54, 55, 56, 57, 58, 59, 83, 84, 85
 86, 87, 88, 89, 90, 91, 92, 93, 94, 95, 96
*Elset, elset=upper, instance=cantilever-1
 91,  92,  93,  94,  95,  96,  97,  98,  99, 100, 196, 197,
198, 199, 200, 201, 202, 203, 204, 205, 206, 207, 208, 209,
210
*End Assembly
**
** MATERIALS
**
*Material, name=material1
*Include, input=material1.inp
*Material, name=material2
*Include, input=material2.inp
**
** BOUNDARY CONDITIONS
**
** Name: clamped Type: Displacement/Rotation
*Boundary
_PickedSet37, 1, 1
_PickedSet37, 2, 2
** ----------------------------------------------------------------
**
** STEP: load
**
*Step, name=load, nlgeom=YES
*Static
0.1, 1., 1e-05, 1.
**
** LOADS
**
** Name: force   Type: Concentrated force
*Cload
```

```
_PickedSet38, 2, -100.
**
** OUTPUT REQUESTS
**
*Restart, write, frequency=0
**
** FIELD OUTPUT: F-Output-1
**
*Output, field, variable=PRESELECT
**
** HISTORY OUTPUT: H-Output-1
**
*Output, history, variable=PRESELECT
* NODE FILE,nset=upper ,frequency=99999
COORD
*FILE FORMAT, ASCII
*End Step
```

As addition to this file, there are also input files `nodes.inp` with nodal coordinates, `elements.inp` with connectivity matrix, and two input files for the material information that have the following simple form:

`material1.inp`

```
*Elastic
20000, 0.3
```

`material2.inp`

```
*Elastic
5000, 0.3
```

With these input files a FE model with mesh visualized in Fig. 4.3 is constructed. The load is applied in one step and is represented by a concentrated vertical fore of the intensity 100 N directed downwards. As a structural response to this load a deformed shape of upper edge is taken. For this purpose a node-set that collects these nodes is created with the name *upper*. The deformed coordinates of these nodes are further written in the external ASCII file. This is achieved by adding following lines at the end of input file:

```
* NODE FILE,nset=upper ,frequency=99999
COORD
*FILE FORMAT, ASCII
```

This command tells ABAQUS to save coordinates of the node-set with the name *upper*, into the external file that should be in ASCII format. Frequency of writing refers to the increments within the step. As in our case we actually need this information only at the end of the step, a large frequency number is given in order to ensure that it will be written only once.

In the present context, all the information that are not changeable within inverse analysis procedure are in the main input file and in the other two that provide information about nodal coordinates and element connectivity. The part that should be continuously changed during the optimization procedure is put into the two input files `material1.inp` and `material2.inp`.

With this structuring, an easy to manipulate FE model is created that will be further implemented within a MATLAB discrepancy function code that should compare computed response to the experimental one. In order to write this routine, first the results from ASCII file written by ABAQUS need to be extracted. For this purpose an additional MATLAB routine is written.

4.1.2 Reading Results from "dot-fil" File

ABAQUS has an option to write all the results from the computation into a human readable ASCII file, with the extension *fil*. In the present context this is a very valuable feature as it represents an easy way to transfer the computed system response into a form which can be used within MATLAB to build a discrepancy function.

The following MATLAB function can be used to read deformed coordinates from the result file that is specified as argument of the function.

```
*************************************************************
function [nodes]=readfil1(file)
% Routine for reading *.fil file from ABAQUS in ASCII format.
% It gets back matrix of deformed coordinates
format long
tic
nodes=[];
%~~~~~~~~~~~~~~~~~~~~~~~~~~~~~~~~~~~~~~~~~~~~~~~~~~~~~
% Reading .fil file
f_fil=fopen(file);
s1=textscan(f_fil,'%c');  % Reading will get back one string
fclose(f_fil);            % data, without any spaces in
s=s1{1}';                 % between
clear s1;
sizeS1=size(s);
sizeS=sizeS1(1,2); % Size of character vector
%~~~~~~~~~~~~~~~~~~~~~~~~~~~~~~~~~~~~~~~~~~~~~~~~~~~~~
% Specifying key words of interest in ABAQUS .fil file
coord_string='I15I3107I'; % Code for coordinates
incr_string='I223I42000D'; % Code for increament
%~~~~~~~~~~~~~~~~~~~~~~~~~~~~~~~~~~~~~~~~~~~~~~~~~~~~~
% Finding places of interest within .fil file
INCRM=strfind(s,incr_string);
firstINC=min(INCRM);
num=firstINC;
nod=0;
%~~~~~~~~~~~~~~~~~~~~~~~~~~~~~~~~~~~~~~~~~~~~~~~~~~~~
% Writing coordinates into NODES matrix
while num<sizeS-8
    if s(num:num+8)==coord_string;
        digit=str2num(s(num+9));
        nod=nod+1;
        nodes(nod,1)=str2num(s(num+10:num+10+digit-1));
        num=num+8+1+digit+1+1;
        foundStar=0;
        coordnum=1;
        while 1-foundStar
            foundD=0;
            begg=num;
            while 2-foundD-foundStar
                if s(num)=='D'
                    foundD=foundD+1;
                end
                if s(num)=='*'
                    foundStar=foundStar+1;
                end
                num=num+1;
            end
            endd=num-2;
```

```
                coordnum=coordnum+1;
                nodes(nod,coordnum)=str2num(s(begg:endd));
            end
        else
        num=num+1;
        end
end
toc
***********************************************************
```

This routine loads an ASCII file as a string data into MATLAB surrounding. Since this file contains a lot of other information that are not of any use for the present purpose, the goal of the above listed routine is to extract the needed data. After the ABAQUS analysis is executed, there will be a file with the same name as the name of "job file" with an extension *fil*. For the example given in this case study, the resulting file has more than 400 lines. The information we need considers coordinates, and these are placed after the code `I 15I 3107I`. This code is somewhat general, in a sense that number at the end, namely 3107, indicates that the information which follows concerns coordinates. Number 15 means that it is a 2D model. For example code `I 16I 3107I` indicates that the data that follow will be a 3D coordinates. With this information it is easy to modify above given routine to be used also for 3D models. Codes of all data which can be written in the resulting dot-fil file can be found in ABAQUS help [1]. Note that within specified options used for command `textscan` in MATALB routine, the string will be loaded without any spaces in between. Therefore, also the code for the coordinates is defined in the same way.

Another important code that is used in the above MATLAB routine is `I 223I 42000D`. This code indicates the beginning of the increment for which the results are written. Since dot-fil file from the beginning contains a lot of information that are not of interest for our purpose, routine is written in the way that it immediately jumps over the first part and positions itself at the beginning of the required data before the process of extracting the coordinates starts. This strategy is very important for complicated models that may produce large dot-fil files. With this approach extracting of data is much shorter.

Finally the cycle of extracting coordinates starts, and the matrix `nodes` is formed that is given back as a result of MATLAB function. For 2D case this will be a $[N_N \times 3]$ matrix, where N_N is the number of nodes for which the coordinates are written. Each line of matrix has as inputs first the number of node for which the coordinates are given, followed by coordinate values (X and Y respectively).

4.1.3 Building Discrepancy Function

After the numerical model is written in the form ready to be easily modified, and by having a MATLAB function for extracting the results of interest from the ASCII file, it is possible to put these two things together within one routine which will compute the discrepancy between experimental and numerical data.

The following MATLAB function is written to compute vector of residuals by comparing the experimental results, stored in an external file, with their numerical counter-part, computed by the use of an ABAQUS FE model.

```
*********************************************************
function e=disfunfem(x)
% Function that quantifies the difference between numerically
% computed response and experimental one (placed in txt file)
parE1=x(1)*10000
parE2=x(2)*10000
load exper.txt % Load experimental results
sortedR=sortrows(exper,1);
%~~~~~~~~~~~~~~~~~~~~~~~~~~~~~~~~~~~~~~~~~~~~~~~~~~~~~
% Changing the input files
%~~~~~~~~~~~~~~~~~~~~~~~~~~~~~~~~~~~~~~~~~~~~~~~~~~~~~
f_fil=fopen('material1.inp');
s=fscanf(f_fil,'%c');
fclose(f_fil);
position=strfind(s,'Elastic');
begg=position+9;
num=begg;
endd=1;
while endd<begg
    if s(num)==','
        endd=num-1;
    end
    num=num+1;
end
% Replacing the modulus of elasticity
oldE=s(begg:endd);
putE=num2str(parE1);
s1=strrep(s,oldE,putE);
f_fil=fopen('material1.inp','w');
fprintf(f_fil,s1);
fclose(f_fil);
% Second input file
f_fil=fopen('material2.inp');
s=fscanf(f_fil,'%c');
fclose(f_fil);
position=strfind(s,'Elastic');
begg=position+9;
num=begg;
endd=1;
while endd<begg
    if s(num)==','
        endd=num-1;
    end
    num=num+1;
end
% Replacing the modulus of elasticity
oldE=s(begg:endd);
putE=num2str(parE2);
s1=strrep(s,oldE,putE);
f_fil=fopen('material2.inp','w');
fprintf(f_fil,s1);
```

```
fclose(f_fil);
%~~~~~~~~~~~~~~~~~~~~~~~~~~~~~~~~~~~~~~~~~~~~~~~~~~~~~~~~
% ABAQUS run
%~~~~~~~~~~~~~~~~~~~~~~~~~~~~~~~~~~~~~~~~~~~~~~~~~~~~~~~~
! abaqus j=cant1_inp interactive
%~~~~~~~~~~~~~~~~~~~~~~~~~~~~~~~~~~~~~~~~~~~~~~~~~~~~~~~~
% Reading .fil file
%~~~~~~~~~~~~~~~~~~~~~~~~~~~~~~~~~~~~~~~~~~~~~~~~~~~~~~~~
[nodes]=readfil1('cant1_inp.fil');
def=nodes(:,2:3);
sorted=sortrows(def,1);
%~~~~~~~~~~~~~~~~~~~~~~~~~~~~~~~~~~~~~~~~~~~~~~~~~~~~~~~~
% Calculating the difference between the deformed shapes
%~~~~~~~~~~~~~~~~~~~~~~~~~~~~~~~~~~~~~~~~~~~~~~~~~~~~~~~~
minC=sorted(1,1);
minR=sortedR(1,1);
minABS=max(minC,minR);
maxC=max(sorted(:,1));
maxR=max(sortedR(:,1));
maxABS=min(maxC,maxR);
NOP=100;
intPOINTS=minABS:(maxABS-minABS)/NOP:maxABS;
calR=interp1(sortedR(:,1),sortedR(:,2),intPOINTS);
calC=interp1(sorted(:,1),sorted(:,2),intPOINTS);
for i=1:size(calR,2)
    e(i,1)=(calR(i)-calC(i));
end
%~~~~~~~~~~~~~~~~~~~~~~~~~~~~~~~~~~~~~~~~~~~~~~~~~~~~~~~~
**********************************************************
```

As an input to the above listed MATLAB function a vector of normalized parameters **x** is given. This vector has the length of two, and its inputs represent normalized values of elasticity moduli of the two materials. The normalization is a useful practice, as then perturbations for the computation of derivatives can be kept rather small like in previous examples. For this case parameter value 1 means 10,000 MPa, as it can be seen in the first two lines of the code. It means that the perturbation of 1E-4 will result in the changes in modulus of elasticity of 1 MPa. Of course without normalization the perturbations should be performed with larger numbers, as it is not expectable to have some difference in the model response if the modulus of elasticity changes by 1E-4 MPa.

The normalization is of special importance when the sought parameters represent different physical properties. Let us imagine the case in which the two sought parameters are Young's modulus and yield limit of some steel. Both quantities are measured by the same unit (i.e. Pa) but there is a difference of three orders of magnitude between them. Performing sensitivity analysis without normalization, and adopting the perturbation value to be the same and equal to, say 1 MPa, will result in much larger sensitivity of measurable quantities to the changes in yield limit than to the changes in elastic modulus since this perturbation represents approximately 1% of the nominal value of the former, while it is about 0.001% of the latter. Such a small change is practically unnoticeable on the measurable quantities resulting in false message that the measurable quantities are not sensitive to the changes of this parameter. Therefore it is very important to normalize sought parameters, and usual practice is to do it in a way that their expected target values are approximately equal to 1.

Within the first few lines input parameters are transferred to their values in MPa (i.e. multiplying them by 10,000), and the experimental data are loaded, which are stored in an external file with the name `exper.txt` represented by a matrix of X-Y coordinates. It is irrelevant if the number of points within this matrix matches or not the number of nodes of the FE mesh, as the code will anyhow perform interpolation when it computes the difference between the two responses. This issue will be commented in more details in the following pages.

The rest of the code is divided into three major parts. First part performs necessary changes on the two input files, namely `material1.inp` and `material2.inp`. Already existing input files are first loaded as strings into MATLAB and 'old' values of Young's moduli are replace by the 'new' ones, that are given as inputs to the MATLAB function. After these changes, new versions of input files are saved to the hard drive.

Second part of the code performs FE simulation by calling the ABAQUS using the command:

```
! abaqus j = cant1_inp interactive
```

Word interactive at the end means that the further execution of MATLAB code will be suspended until FE analysis is not finished. Once the analyses is finished, code proceeds by reading the results from the dot-fil file using the routine `readfil1.m` given in previous section, which is returning as a result matrix of deformed coordinates of the node-set defining the upper surface (i.e. line) of the cantilever model (see Fig. 4.2).

The third and final part of the code is computing the difference between the two responses. The result of this comparison is a vector of residuals which represents the difference between Y-coordinate of the two curves over a certain grid of equidistant points. Figure 4.4 shows an example of the two responses (numerical and experimental one) and the way the vector of residuals is constructed over a given fixed grid along X-axis.

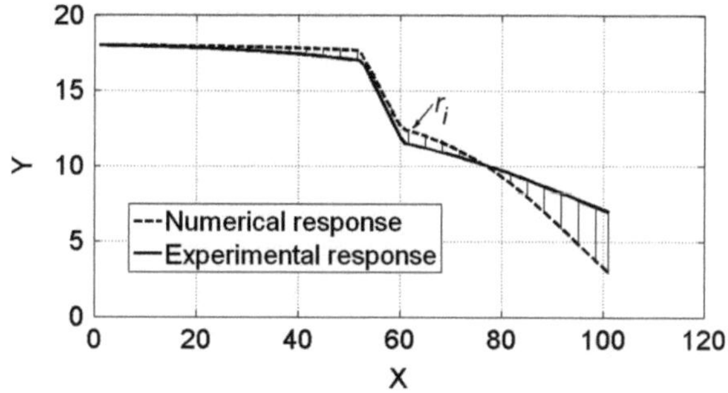

Fig. 4.4 Residuals for discrepancy function that quantifies the difference between the two curves

The number of points along which the coordinates are computed is fixed in the code by the variable `NOP`. The corresponding Y-coordinates are then interpolated from both resources of data (i.e. experimental and computed) in order to have the amplitudes at exactly the same abscise. This interpolation takes also care that the grid of points should be placed over the range in which both of the curves exist.

Before proceeding to the final step in which an optimization algorithm will be constructed it is needed also to have some representation of experimental results to feed the discrepancy function. At this stage it can be done with pseudo-experimental data, as mentioned earlier in Chap. 1. For this purpose, a FE simulation can be executed once by attributing to the parameters some values, say in this case for Young's modulus of material 1 value 20,000 MPa, and for material 2 value 5,000 MPa, and resulting deformed shape of the upper edge can be saved in file `exper.txt`.

After performing this step a MATALB code that computes discrepancy between the two responses for the experiment studied here is fully prepared and we can proceed by writing a code to solve the optimization problem. It should be mentioned that a useful practice is to use a different numerical model to generate pseudo-experimental data (e.g. by changing FE mesh of the model). With this approach target parameter values with the numerical model used in the inverse analysis procedure will not produce exactly the same results as pseudo-experimental one, which is always the case when dealing with the real experimental data.

4.1.4 Solving the Optimization Problem

In order to solve the resulting optimization problem we will use a dog-leg trust region algorithm presented in Chap. 2. Considering that the discrepancy function is in least squares form we will make a use of approximated Hessian matrix in order to avoid numerous FE simulations required to compute second derivatives. However, in order to keep the generality of the code, an option to use full Hessian matrix is implemented, and at the beginning of the program user selects together with other options which one will be used.

In what follows a listing of MATLAB code that minimizes previously created discrepancy function by a trust region dog-leg method is given.

```
*************************************************************
% Trust region algorithm with dog-leg approach for sub-problem
clear
clc
%~~~~~~~~~~~~~~~~~~~~~~~~~~~~~~~~~~~~~~~~~~~~~~~~~~~~~~~~~~~
% Setting the options
minchg=1e-5; % Minimum change in parameters
MAXIT=30; % Maximum allowed number of iterations
guess=[1.5;1.5]; % Initial guess of parameters
pert=1e-4; % Perturbation for the first derivatives
res=10;
TRrad=0.8; % Initial Trust Region radius
HessMod=1; % Indication for modification of Hessian
Hessapp=1; % Indication for Hessian approximation
mingrad=1e-6; % Gradient value at which to terminate optim.
%~~~~~~~~~~~~~~~~~~~~~~~~~~~~~~~~~~~~~~~~~~~~~~~~~~~~~~~~~~
% Computing for the first time value of function
iter=0;
eV=disfunfem(guess);
e0=0.5*eV'*eV;
%~~~~~~~~~~~~~~~~~~~~~~~~~~~~~~~~~~~~~~~~~~~~~~~~~~~~~~~~~~
% Optimization cycle
while res>1e-6
itiner(iter+1,1:2)=guess';
itiner(iter+1,3)=e0;
iter=iter+1;
if Hessapp==1
    [HESS,grad]=comhessapp(@disfunfem,guess,pert,eV);
else
    [HESS,grad]=comhess(@disfunfem,guess,pert,eV);
end
% Checking the gradient
if grad'*grad<mingrad
    break
end
% Ensuring that Hessian is positive-definite
if HessMod==1
lambdas=eigs(HESS);
if lambdas(1)>0 && lambdas(2)>0
    hessmod=0;
else
    coeff=mean(abs(lambdas));
    posdef=0;
    while posdef<1
        HESSm=HESS+coeff*eye(2);
        hessmod=1;
        lmb=eigs(HESSm);
        if lmb(1)>0 && lmb(2)>0
            posdef=1;
        else
            coeff=coeff*1.5;
```

```
            end
        end
    end
else
    hessmod=0;
end
stpdsc=-grad/norm(grad);
if hessmod==0
    newton=-inv(HESS)*grad;
else
    newton=-inv(HESSm)*grad;
end
%~~~~~~~~~~~~~~~~~~~~~~~~~~~~~~~~~~~~~~~~~~~~~~~~~~~~~
% Computing step
accepted=0;
rejected=0;
while accepted<1
if norm(newton)<TRrad
    pDL=newton; % Dog Leg step
else
    % Finding the Cauchy point
    pc=cauchypnt(e0,stpdsc,grad,HESS,TRrad);
    % Finding minimizer within trust region (Dog Leg step)
    diff=newton-pc;
    dimV=size(newton,1);
    cf=[0,0,-TRrad^2];
    for ii=1:dimV
        cf(1)=cf(1)+diff(ii)^2;
        cf(2)=cf(2)+2*pc(ii)*diff(ii);
        cf(3)=cf(3)+pc(ii)^2;
    end
    alfa=max(roots(cf)); % Taking the positive root
    pDL=pc+alfa*diff;
end
    predred=-(pDL'*grad+0.5*pDL'*HESS*pDL);
    guess1=guess+pDL; % Next iterate
    eV=disfunfem(guess1);
    e1tr=0.5*eV'*eV;
    realred=e0-e1tr;
    ratio=realred/predred;
    if ratio<0
        TRrad=TRrad/1.2;
        rejected=rejected+1;
        disp('Step is rejected!!!')
    else
        accepted=1;
        itiner(iter+1,1:2)=guess1';
        itiner(iter+1,3)=e1tr;
        if ratio<0.2
            TRrad=TRrad/1.2;
        end
        if ratio>0.6
            TRrad=TRrad*1.2;
        end
    end
```

```
    if rejected==3
        accepted=1;
        guess1=guess; % Since there was no improvement
    end
end
if rejected==3
    break
end
iterscr=['Iteration: ',num2str(iter)];
disp(iterscr)
dfscr=['Current value of objective function: ',num2str(e1tr)];
disp(dfscr)
guess=guess1;
res=e1tr;
e0=e1tr;
if iter>MAXIT % Terminate after iterations reach MAXIT
    res=0;
end
% Checking convergence options
if abs(itiner(iter+1,1)-itiner(iter,1))<minchg ||
abs(itiner(iter+1,2)-itiner(iter,2))<minchg;
    res=0;
end
end
***********************************************************

***********************************************************
function [HESS,grad]=comhessapp(FUNNAME,point,pert,e0)
% Computing the Approximated Hessian matrix
eS0=0.5*e0'*e0;
% Computing Jacobian
for i=1:size(point,1)
    pointp=point;
    pointp(i)=pointp(i)+pert;
    e1=FUNNAME(pointp);
    eS1=0.5*e1'*e1;
    grad(i,1)=(eS1-eS0)/pert;
    J(:,i)=(e1-e0)/pert;
end
HESS=J'*J;
***********************************************************

***********************************************************
function [HESS,grad]=comhess(FUNNAME,point,pert,eV)
% Computing the Hessian matrix
e0=0.5*eV'*eV;
for i=1:size(point,1)
    pointp=point;
    pointp(i)=pointp(i)+pert;
    eV=FUNNAME(pointp);
    e1=0.5*eV'*eV;
    grad(i,1)=(e1-e0)/pert;
    pointp(i)=pointp(i)+pert;
    eV=FUNNAME(pointp);
```

```
    e2=0.5*eV'*eV;
    HESS(i,i)=(e0-2*e1+e2)/(pert^2);
end
% mixed derivative
for i=1:size(point,1)-1
    for j=2:size(point,1)
        pointp=point;
        pointp(i)=pointp(i)+pert;
        eV=FUNNAME(pointp);
        e1=0.5*eV'*eV;
        pointp=point;
        pointp(j)=pointp(j)+pert;
        eV=FUNNAME(pointp);
        e2=0.5*eV'*eV;
        pointp=point;
        pointp(i)=pointp(i)+pert;
        pointp(j)=pointp(j)+pert;
        eV=FUNNAME(pointp);
        e11=0.5*eV'*eV;
        HESS(i,j)=1/pert*((e11-e1)/pert-(e2-e0)/pert);
        HESS(j,i)=HESS(i,j);
    end
end
***********************************************************
```

The above listing includes three different routines. The first one is the main program that solves the inverse problem, while the remaining two are MATLAB functions used to compute full Hessian matrix, or approximated one. These latter two are slightly modified with respect to those listed in Chap. 2, in order to reduce the number of simulations performed. Therefore, both of the functions are receiving as inputs vector of residuals for the current point, and so just the analyses for perturbed parameters are computed. Further, both of the functions are returning as a result also gradient vector together with Hessian matrix, so it is not computed anymore within the main program of trust region optimization. As addition to these two MATLAB functions, program requires also function `disfunfem` listed in previous section, and function `cauchypnt` for Cauchy point solution listed in Chap. 2.

The optimization algorithm is very similar to the one presented in Chap. 2. At the beginning of the program, where the main options are set, it is also added the one that prescribes the value of gradient at which the optimization should be terminated (i.e. to be considered as zero-gradient). It represents an additional stopping criterion added to those previously discussed in the examples listed in Chap. 2. Result of the optimization is stored in `itiner` matrix where each line collects values of parameters together with the corresponding value of the objective function.

4.1.5 Results of Inverse Analyses

After the whole inverse analysis procedure is designed it is time to test it using pseudo-experimental data. As mentioned above, pseudo-experimental data are produced for the parameter values of $E_1 = 20{,}000$ MPa and $E_2 = 5{,}000$ MPa.

Keeping the parameters normalized around value 1 (i.e. target value for the first Young's modulus is 2, and for the second one is 0.5) perturbation of 1E-4 turned out to be a good value which produces a reliable response. In some more complex objective functions (usually in optimization problems with larger number of parameters) the selection of this parameter may become more important issue as then the discrepancy function is usually less smooth.

The inverse analysis should be solved different times starting from various initialization points. Figure 4.5 visualizes result of one optimization starting from inverted values of parameters (i.e. attributing to E_1 target value of E_2 and vice versa). The convergence was quit fast and already after five iterations the parameters were close to their target values. Optimization was terminated after the gradient was smaller than the value specified within the options. Initial value of trust region radius was set to 0.8, and it turned out to be rather large as it led to the

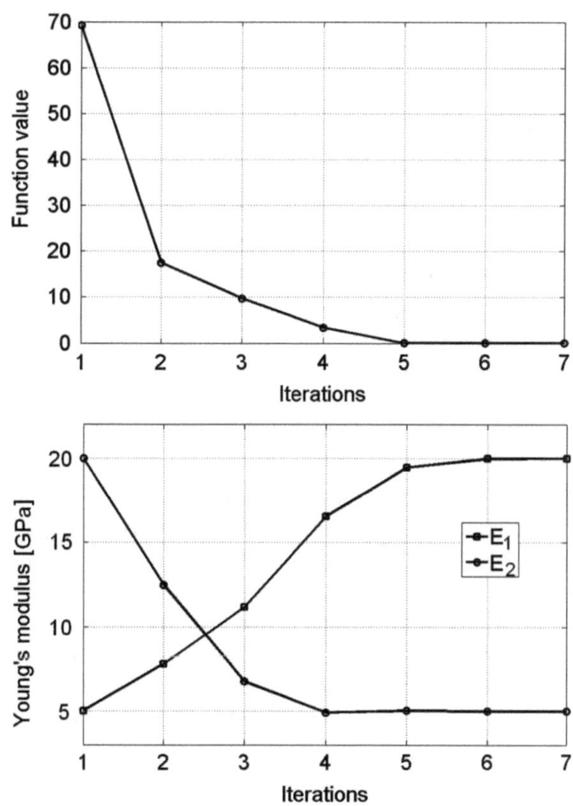

Fig. 4.5 Result of optimization: initialization 1 – reduction of the objective function (*up*) and changes of parameters through iterations (*down*)

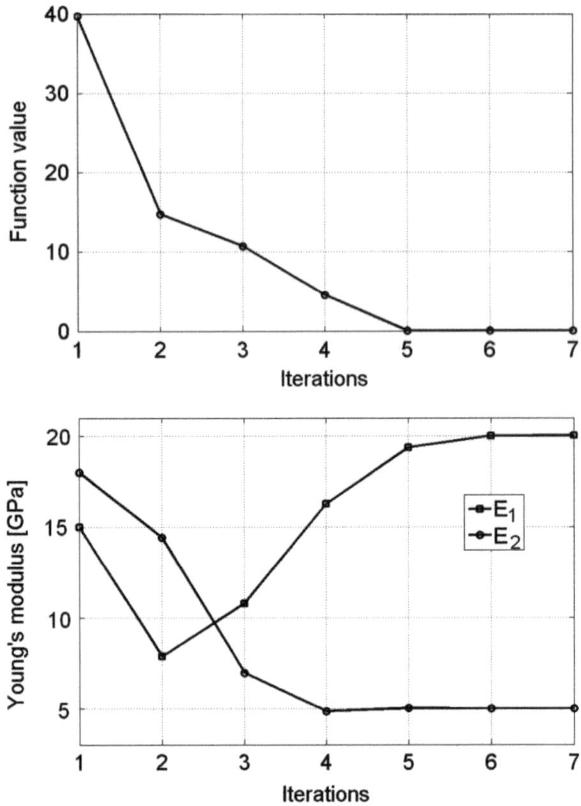

Fig. 4.6 Result of optimization: initialization 2 – reduction of the objective function (*up*) and changes of parameters through iterations (*down*)

rejection of the step in iteration 2. After this rejection, the trust region was reduced according to the algorithm and there were no further step rejections up to the convergence.

Figure 4.6 visualizes in the same manner optimization results for different initialization point. Also in this case the optimization was terminated after seven iterations, when gradient reached value smaller than the one prescribed by the options. Converged parameters were practically matching target values in both optimizations.

Since the first optimization resulted in one step rejection, quit "early" in the optimization (i.e. in the second iteration), it suggested that its value was a bit too-large. In fact, by attributing to the initial trust region radius value of 0.4 in the second initialization, the optimization was fully executed without any step rejection. Although this value is not crucial from the global convergence point of view, it is affecting slightly the overall performance of the optimization algorithm as it may increase the total number of analysis involved.

As announced in Chap. 1, once that the whole procedure is designed it is highly desirable to make more profound tests. Apart of the analyses with results visualized in Figs. 4.5 and 4.6 it is also useful to perform further numerical exercises by adding an artificial noise to the pseudo-experimental data, in order to understand how much measuring error is influencing the accuracy of the measurements. Finally, the same computation should be performed also for different parameter combination used as target values in order to verify that the problem is well-posed not only in some parameter range. This part is left to the reader to be done as an exercise.

4.2 Case Study 2: Assessment of Plastic Parameters of Thin Plate

In the second case study we will consider a thin steel plate with anisotropic plastic behavior. Sheet metal forming, that is frequently used e.g. in car industry, usually produces plates with anisotropic properties. In order to model this behavior often Hill's yield criterion is used (see e.g. [2, 3]). In the case of orthotropic sheets this criterion has the following form

$$f_h = (G+H)\sigma_{11}^2 + (F+H)\sigma_{22}^2 - 2H\sigma_{11}\sigma_{22} + 2N\sigma_{12}^2 - 1 = 0 \tag{4.1}$$

where F, G, H and N are anisotropic constants, and σ_{11}, σ_{22} and σ_{12} are stress components in the plane of the sheet. Traditionally reference system is adopted such that the axis 1 coincides with rolling direction, axis 2 is perpendicular in the plane of the sheet, and axis 3 is through-thickness direction.

Above anisotropic constants are usually expressed in terms of yield limits corresponding to different directions by the following relations

$$F = \frac{1}{2}\left(\frac{1}{R_{22}^2} + \frac{1}{R_{33}^2} - \frac{1}{R_{11}^2}\right) \tag{4.2a}$$

$$G = \frac{1}{2}\left(\frac{1}{R_{33}^2} + \frac{1}{R_{11}^2} - \frac{1}{R_{22}^2}\right) \tag{4.2b}$$

$$H = \frac{1}{2}\left(\frac{1}{R_{11}^2} + \frac{1}{R_{22}^2} - \frac{1}{R_{33}^2}\right) \tag{4.2c}$$

$$N = \frac{3}{2R_{12}^2} \tag{4.2d}$$

where R_{ij} are anisotropic yield stress ratios, namely

$$R_{11} = \frac{\bar{\sigma}_{11}}{\sigma_0}; R_{22} = \frac{\bar{\sigma}_{22}}{\sigma_0}; R_{33} = \frac{\bar{\sigma}_{33}}{\sigma_0}; R_{12} = \frac{\bar{\sigma}_{12}}{\tau_0}; \text{ with } \tau_0 = \frac{\sigma_0}{\sqrt{3}} \tag{4.3}$$

With the above definition, Hill's yield surface for the plane stress case is defined by one nonzero stress component σ_0 and additional two ratios (i.e. R_{11} can be set to 1 and so the yielding for this direction is directly given by σ_0, and assuming the same value for through thickness direction, namely $R_{33} = R_{11}$), allowing therefore to have three different values for yielding stresses for direction 1, direction 2 and shearing.

As addition to these three parameters also hardening can be considered. The simplest approach of introducing hardening to an anisotropic yield criterion is to assume that the yield surface doesn't change in shape as the material hardens. Assuming an exponential hardening, the above scalar value σ_0 is not constant anymore, but an exponential function of equivalent plastic strain, namely

$$\sigma_0 = \left(\frac{E\varepsilon_p^{eq}}{\sigma^Y}\right)^n \tag{4.4}$$

with n being an exponent of hardening, and ε_{eq}computed according to Hill's yield criterion (see e.g. [3]).

As the plastic deformation evolves, yield surface expands, by changing the yield limit in direction 1 according to Eq. 4.4, while keeping all the ratios (4.3) unchanged, preserving therefore the shape as previously mentioned.

With this definition, plastic behavior for thin plate material is defined by four constants: three yield limits (direction 1, direction 2 and shearing), and one exponent of hardening.

In order to calibrate these constants a biaxial tests can be used (see e.g. [4, 5]). Sometimes a circular hole is introduce to the plate sample, or a cruciform shape is used (or both cruciform with a circular hole) in order to make the distribution of the stresses less uniform, and therefore the experimentally measured response more sensitive to the variation of the material parameters.

For this case study let us consider a simple thin plate with the circular hole subjected to a biaxial test (see Fig. 4.7). Let us imagine that the objective is to design an inverse analyses procedure that will use a biaxial test in order to assess four plastic parameters: three yield limits and one exponent of hardening. Let us further assume that elastic parameters are known in advanced and that the material in elasticity is isotropic, defined therefore by two constants: one elastic modulus and Poisson's ratio.

As an experimental data two curves will be taken: applied force versus obtained displacement for horizontal direction, and the same for vertical direction. This is the most traditional approach, as most of the machines that are used for the biaxial test are supplying this information. Additionally, also the displacement field can be used e.g. provided by the use of Digital Image Correlation (see e.g. [6]), but it requires additional equipment to be involved in the experiment.

After the experiment is adopted we will proceed with the second step of the procedure, namely with building of a numerical model of it.

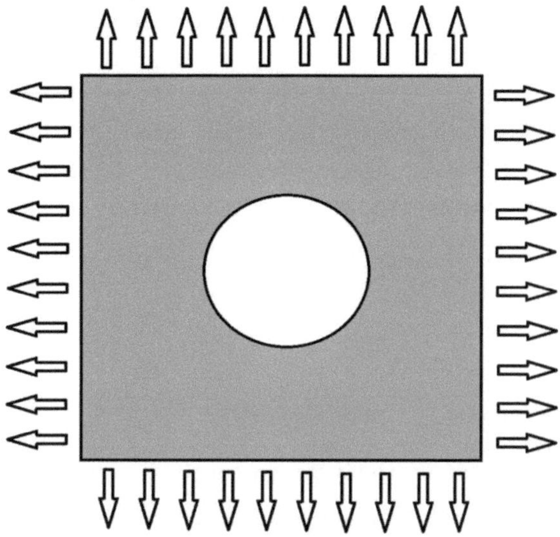

Fig. 4.7 Thin plate with circular hole subjected to biaxial test

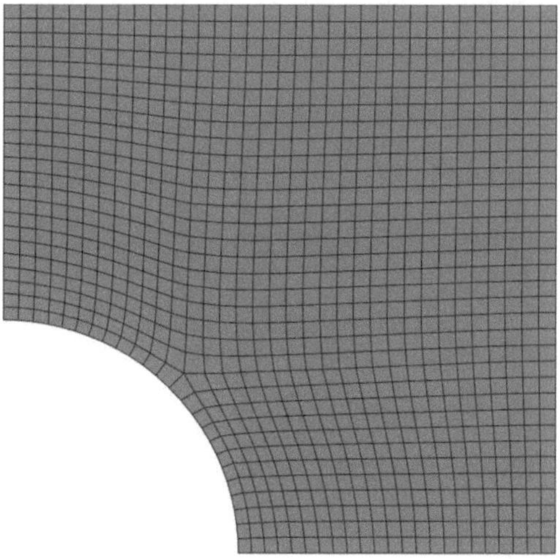

Fig. 4.8 FE model of one-quarter of thin plate with circular hole

4.2.1 FE Model of the Experiment

Like in previous case, also for this case study we will use a commercial code ABAQUS in order to simulate the experiment. In order to reduce computing times, we will make a use of symmetry and build therefore only one quarter of the specimen, applying symmetry conditions along cut lines. The model is 2D plane stress and the adopted mesh is visualized in Fig. 4.8.

In the same manner as previously, input file that defines numerical model will be divided in couple of files. The only file that needs to be changed is the one that gives the information about material, namely `material1.inp`. The main input file is given below

```
*Heading
** Job name: bi_ax1 Model name: Model-1
*Preprint, echo=NO, model=NO, history=NO, contact=NO
**
** PARTS
**
*Part, name=plate
*End Part
**
**
** ASSEMBLY
**
*Assembly, name=Assembly
**
*Instance, name=plate-1, part=plate
*Node
*Include, input=nodes.inp
*Element, type=CPS4
*Include, input=elements.inp
*Nset, nset=_PickedSet8, internal, generate
    1,  1106,     1
*Elset, elset=_PickedSet8, internal, generate
    1,  1036,     1
*Nset, nset=_PickedSet9, internal, generate
    1,  1106,     1
*Elset, elset=_PickedSet9, internal, generate
    1,  1036,     1
*Orientation, name=Ori-3
1.,       0.,        0.,       0.,        1.,       0.
1, 0.
** Region: (solid:Picked), (Material Orientation:Picked)
*Elset, elset=_PickedSet8, internal, generate
    1,  1036,     1
** Section: solid
*Solid    Section,    elset=_PickedSet8,    orientation=Ori-3,
material=steel
1.,
*End Instance
**
*Nset, nset=_PickedSet6, internal, instance=plate-1
   7,   8,   9,  77,  78,  79,  80,  81,  82,  83,  84,  85,
86,  87, 109, 110, 111, 112, 113, 114, 115, 116, 117, 118,
119, 120, 121, 122, 123, 124, 125, 126, 127, 128, 129
*Elset, elset=_PickedSet6, internal, instance=plate-1
  46,  68,  90, 112, 134, 156, 178, 200, 222, 244, 266, 288,
289, 290, 291, 292, 293, 294, 295, 296, 297, 298, 299, 300,
301, 302, 303, 304, 305, 306, 307, 308, 309, 310
*Nset, nset=_PickedSet12, internal, instance=plate-1
   9,  10,  11, 130, 131, 132, 133, 134, 135, 136, 137, 138,
139, 140, 141, 142, 143, 144, 145, 146, 147, 148, 149, 150,
172, 173, 174, 175, 176, 177, 178, 179, 180, 181, 182
```

```
*Elset, elset=_PickedSet12, internal, instance=plate-1
 310, 332, 354, 376, 398, 420, 442, 464, 486, 508, 530, 552,
574, 596, 618, 640, 662, 684, 706, 728, 750, 772, 773, 774,
775, 776, 777, 778, 779, 780, 781, 782, 783, 784
*Nset, nset=_PickedSet19, internal, instance=plate-1
   1,    2,   11, 183, 184, 185, 186, 187, 188, 189, 190, 191,
192, 193, 194, 195, 196, 197, 198, 199, 200, 201, 202, 203
*Elset, elset=_PickedSet19, internal, instance=plate-1
    1, 784, 796, 808, 820, 832, 844, 856, 868, 880, 892, 904,
916, 928, 940, 952, 964, 976, 988, 1000, 1012, 1024, 1036
*Nset, nset=_PickedSet20, internal, instance=plate-1
  5,   6,   7, 56, 57, 58, 59, 60, 61, 62, 63, 64, 65, 66, 67,
68, 69, 70, 71, 72, 73, 74, 75, 76
*Elset,   elset=_PickedSet20,   internal,   instance=plate-1,
generate
 24,  46,   1
*Nset, nset=bottom, instance=plate-1
  5,   6,   7, 56, 57, 58, 59, 60, 61, 62, 63, 64, 65, 66, 67,
68, 69, 70, 71, 72, 73, 74, 75, 76
*Elset, elset=bottom, instance=plate-1, generate
 24,  46,   1
*Nset, nset=left, instance=plate-1
   1,    2,   11, 183, 184, 185, 186, 187, 188, 189, 190, 191,
192, 193, 194, 195, 196, 197, 198, 199, 200, 201, 202, 203
*Elset, elset=left, instance=plate-1
    1, 784, 796, 808, 820, 832, 844, 856, 868, 880, 892, 904,
916, 928, 940, 952, 964, 976, 988, 1000, 1012, 1024, 1036
*Elset,  elset=__PickedSurf22_S2,  internal,  instance=plate-1,
generate
 310,  772,   22
*Elset,  elset=__PickedSurf22_S1,  internal,  instance=plate-1,
generate
 773,  784,    1
*Surface, type=ELEMENT, name=_PickedSurf22, internal
__PickedSurf22_S2, S2
__PickedSurf22_S1, S1
*Elset,  elset=__PickedSurf23_S2,  internal,  instance=plate-1,
generate
  46,  288,   22
*Elset,  elset=__PickedSurf23_S1,  internal,  instance=plate-1,
generate
 289,  310,    1
*Surface, type=ELEMENT, name=_PickedSurf23, internal
__PickedSurf23_S2, S2
__PickedSurf23_S1, S1
*End Assembly
**
** MATERIALS
**
*Material, name=steel
*Elastic
```

```
200000., 0.3
*Include, input=material.inp
** ----------------------------------------------------------------
**
** STEP: load
**
*Step, name=load, nlgeom=YES
*Static
0.02, 1., 1e-05, 0.05
**
** BOUNDARY CONDITIONS
**
** Name: bottom Type: Displacement/Rotation
*Boundary
_PickedSet20, 2, 2
** Name: left Type: Displacement/Rotation
*Boundary
_PickedSet19, 1, 1
** Name: rigth Type: Displacement/Rotation
*Boundary
_PickedSet6, 1, 1, 0.2
** Name: up Type: Displacement/Rotation
*Boundary
_PickedSet12, 2, 2, 0.2
**
** OUTPUT REQUESTS
**
** FIELD OUTPUT: F-Output-1
**
*Output, field, variable=PRESELECT
**
** HISTORY OUTPUT: H-Output-1
**
*Output, history, variable=PRESELECT
* NODE FILE,nset=left
RF
* NODE FILE,nset=bottom
RF
*FILE FORMAT, ASCII
*End Step
```

The biaxial test is modeled as displacement controlled. Therefore, within one loading step the displacements prescribed to two edges (the upper one and the one on the right) are applied simultaneously. Note that, in order to use anisotropic material model, a local orientation was introduced to the model, with axes 1 coinciding with horizontal one, 2 with vertical, and axis 3 coinciding with a through thickness direction (orientation with name `Ori-3` in the input file).

As in previous case study, additional input files to the main one are `nodes.inp` with nodal coordinates, `elements.inp` with connectivity matrix (both not presented here) and input file for material information that has the following form:

material1.inp

```
*Plastic
400,0
416.5519,0.001
428.7094,0.002
438.3833,0.003
446.4493,0.004
453.3846,0.005
459.4793,0.006
464.9232,0.007
469.8476,0.008
474.3471,0.009
478.4925,0.01
508.3926,0.02
542.3528,0.04
563.8923,0.06
579.8803,0.08
592.6755,0.1
603.3829,0.12
612.6125,0.14
620.7382,0.16
628.0065,0.18
634.1884,0.2
*Potential
1, 0.8,1,0.99593,1,1
```

Separate input file for the material information consists only of changeable data. Since for this case study elastic parameters are known, the information about them remained in the main input file, to keep therefore changeable file as small as possible. This should be a general practice when preparing ABAQUS input files for inverse analysis.

Hill yield criterion is introduced in ABAQUS by the following command:

`*Potential`

which is followed by ratios given in Eq. 4.3. All six ratios need to be specified also for 2D models, like plane stress in our case. To other ratios that are not playing any role is attributed value 1, and are not changed throughout the inverse analyses procedure. Apart of ratios, as the model has also hardening, the relationship between σ_0 and equivalent plastic strain needs to be supplied, and this part is done in the tabular form, right after the command

`*Plastic`

giving first the information about stress and then about equivalent plastic strain. This table in general can proceed up to a very large values of plastic strain, and ABAQUS assumes perfectly-plastic behavior after the last specified line. For instance, in the example of input file visualized here, for the equivalent plastic strains larger than 0.2 stress will be constant and equal to the last specified value, namely 634.

Since we need to have force-displacement curves as experimental information, unlike the previous example, here we are not only interested in final deformed shape (corresponding to fully applied prescribed loads), but we need also to follow intermediate solutions. Therefore, it is useful to fixed also maximum possible increment during iterative solution. It is done at the beginning of the step with the following commands:

```
*Step, name=load, nlgeom=YES
*Static
0.02, 1., 1e-05, 0.05
```

First number after the word `* Static` gives the value for initial increment size with which ABAQUS will start calculations. However, if during the calculation there are strong indications that the step can be enlarged, ABAQUS will increase it, in order to speed up computation, but not above the last specified number in the line that follows after the word `* Static` (here 0.05). With this constrain we keep the increments small enough in order to have a reasonable number of intermediate equilibrium solutions for construction of force-displacement curves.

Finally at the end of calculation we need to extract these curves, and therefore reaction forces of nodes that are belonging to vertical line of symmetry and horizontal line of symmetry needs to be written down in the external file. This is done in a similar way as in previous case study, namely:

```
* NODE FILE,nset=left
RF
* NODE FILE,nset=bottom
RF
*FILE FORMAT, ASCII
```

Since we need the result after each increment, the frequency of writing is not specified. In such cases, ABAQUS by default writes the result after each increment. The two lines specified above write the reaction forces for two node-sets named "left" and "bottom". Naturally, these needs to be first created which is done within the main input file. Nodes with numbers that follow after the lines:

```
*Nset, nset=bottom, instance=plate-1
     *Nset, nset=left, instance=plate-1
```

represent members of these node-sets.

This completes the preparation of numerical model. Constructed numerical model can be used within the discrepancy function that we will further write. However, since the needed results in this case study are different from those in previous (i.e. here we need force-displacement curves) some modifications of previously given MATLAB function `readfil1.m` need to be implemented.

4.2.2 Reading the Results from "dot-fil" File

For this case study, the data we need to extract from the simulation are two force-displacement curves in two perpendicular directions. The simulation is done under the control of displacements, as we earlier saw, and the resulting reaction forces for the two node-sets of interest are written in the external dot-fil file.

To extract these data from the resulting ASCII file a MATLAB function is used with the listing given below

```
***********************************************************
function [disp,incr]=readfil2(file)
% Routine for reading *.fil file from ABAQUS in ASCII format.
% It gets back vector of increments and matrix of reaction
% forces for every single node for which it is written.
tic
disp=[];
incr=[];
%~~~~~~~~~~~~~~~~~~~~~~~~~~~~~~~~~~~~~~~~~~~~~~~~~~~~~~~~~~~
% Reading .fil file
f_fil=fopen(file);
s1=textscan(f_fil,'%c');   % This reading will get back one
fclose(f_fil);             % string data, without any spaces in
s=s1{1}';                  % between, stored in s
clear s1;
sizeS1=size(s);
sizeS=sizeS1(1,2); % Size of character vector
%~~~~~~~~~~~~~~~~~~~~~~~~~~~~~~~~~~~~~~~~~~~~~~~~~~~~~~~~~~~
% Defining strings codes
force_string='I15I3104I'; % Code for reaction force
incr_string='I223I42000D'; % Code for increment string
%~~~~~~~~~~~~~~~~~~~~~~~~~~~~~~~~~~~~~~~~~~~~~~~~~~~~~~~~~~~
% Finding places of interest within .fil file
INCRM=strfind(s,incr_string);
firstINC=min(INCRM);
num=firstINC;
curve=0;
%~~~~~~~~~~~~~~~~~~~~~~~~~~~~~~~~~~~~~~~~~~~~~~~~~~~~~~~~~~~
% Writing Increment vector
while num<sizeS-10
        if s(num:num+10)==incr_string
        curve=curve+1;
        num=num+11;
        begg=num;
        foundD=0;
        while 2-foundD
            if s(num)=='D'
                foundD=foundD+1;
            end
            num=num+1;
        end
        endd=num-2;
        reldisp=str2num(s(begg:endd));
        incr(curve,1)=reldisp;
    end
    num=num+1;
end
%~~~~~~~~~~~~~~~~~~~~~~~~~~~~~~~~~~~~~~~~~~~~~~~~~~~~~~~~~~~
% Writing displacements into DISP matrix
num=firstINC;
nod=0;
```

```
while num<sizeS-8
    if s(num:num+8)==force_string;
        digit=str2num(s(num+9));
        nod=nod+1;
        disp(nod,1)=str2num(s(num+10:num+10+digit-1));
        num=num+8+1+digit+1+1;
        foundStar=0;
        entnum=1;
        while 1-foundStar
            foundD=0;
            begg=num;
            while 2-foundD-foundStar
                if s(num)=='D'
                    foundD=foundD+1;
                end
                if s(num)=='*'
                    foundStar=foundStar+1;
                end
                num=num+1;
            end
            endd=num-2;
            entnum=entnum+1;
            disp(nod,entnum)=str2num(s(begg:endd));
        end
    else
    num=num+1;
    end
 end
toc
************************************************************
```

Structuring of the MATLAB routine is very similar to the one given in previous case study, with the necessary modifications in order to be adjusted for the problem studied here. Since the simulation is done as displacement controlled and the prescribed displacements (in both directions) are not parameters that are changeable from one simulation to another, it is enough to read from result file just the values of the increments. ABAQUS indicates them with values from 0 to 1, where 1 means application of full prescribed external action (in this case displacement). It is therefore enough to multiple this vector by the prescribed value of displacement at the boundary in order to obtain the history of displacements through the simulation. MATLAB function `readfil2.m` listed above stores information about increments in vector `incr` that is computed as one output of the function.

Second part of the information that we need to extract from dot-fil file considers reaction forces. ABAQUS input file is written in order to store the reaction force for each node belonging to node-sets with the names "left" and "bottom", after every increment. Both reaction components will be written (since `RF` key word is used in the input file, and therefore not any particular component is specified, in such cases, ABAQUS by default writes all the components). The above MATLAB function is written in such way that these results will be stored in the matrix `DISP`, in the order of their appearance. Each line of the matrix has the node number, followed by the two components of reaction, since the problem is two-dimensional. The number of lines of this matrix will be equal to the total number of nodes in both node-sets multiplied by number of increments (i.e. the number of times the reaction forces are written in the file). In order to find the resulting reaction force it is enough to make a summation of all the nodal components belonging to one node set for each increment. This operation will be done inside discrepancy function MATLAB routine.

4.2.3 Building Discrepancy Function

Next step in building of Inverse Analysis procedure is to write a MATLAB discrepancy function that will be later minimized combining previously developed pieces. The goal of this function is to quantify the discrepancy between experimental and computed data, in this case the two force-displacement curves.

Here below is given a listing of MATLAB function that solves this problem

```
************************************************************
function e=disfunfem(x)
% Function that quantifies the difference between numerically
% computed response and experimental one(placed in txt file)
parY=x(1)*400
parY2=x(2)*320
parY12=x(3)*230
parN=x(4)/10
parE=200000; % Young's modulus is unchanged
load exphor.txt % Load experimental curves
load expver.txt
%~~~~~~~~~~~~~~~~~~~~~~~~~~~~~~~~~~~~~~~~~~~~~~~~~~~~~~~~~~~~
% Changing the input file
%~~~~~~~~~~~~~~~~~~~~~~~~~~~~~~~~~~~~~~~~~~~~~~~~~~~~~~~~~~~~
% Preparing the table for stress-strain curve (ssc)
br=0;
elstrain=parY/parE;
for plstrain=0:0.001:0.01
    br=br+1;
    ssc(br,1)=parY^(1-
parN)*parE^parN*(elstrain+plstrain)^parN;
    ssc(br,2)=plstrain;
end
for plstrain=0.02:0.02:0.2
    br=br+1;
    ssc(br,1)=parY^(1-
parN)*parE^parN*(elstrain+plstrain)^parN;
    ssc(br,2)=plstrain;
end
% Preparing potential
R2=parY2/parY;
R4=parY12/(parY/sqrt(3));
% Making changes in the input file
f_fil=fopen('material.inp'); % Loading the input file
s=fscanf(f_fil,'%c');
fclose(f_fil);
position=strfind(s,'Plastic');
linebreak=s(position+7:position+8);
begg=position+9;
num=begg;
newPL=[num2str(ssc(1,1)),',',num2str(ssc(1,2)),linebreak];
for i=2:size(ssc,1)
    line=[num2str(ssc(i,1)),',',num2str(ssc(i,2))];
    newPL=[newPL,line,linebreak];
end
newPOT=['*Potential',linebreak,'1,
',num2str(R2),',1,',num2str(R4),',1,1'];
s1=[s(1:begg-1),newPL,newPOT];
f_fil=fopen('material.inp','w');
fprintf(f_fil,s1);
fclose(f_fil);
```

```
%~~~~~~~~~~~~~~~~~~~~~~~~~~~~~~~~~~~~~~~~~~~~~~~~~~~~~~~~~~~~~~
% ABAQUS run
%~~~~~~~~~~~~~~~~~~~~~~~~~~~~~~~~~~~~~~~~~~~~~~~~~~~~~~~~~~~~~~
% ! abaqus j=bi_ax_inp interactive
%~~~~~~~~~~~~~~~~~~~~~~~~~~~~~~~~~~~~~~~~~~~~~~~~~~~~~~~~~~~~~~
% Reading .fil file
%~~~~~~~~~~~~~~~~~~~~~~~~~~~~~~~~~~~~~~~~~~~~~~~~~~~~~~~~~~~~~~
[disp,indcur]=readfil2('bi_ax_inp.fil');
NINC=size(indcur);
for i=1:NINC
    hor(i,1)=indcur(i)*0.2;
    ver(i,1)=indcur(i)*0.2;
    hor(i,2)=sum(disp(i*48-47:i*48-24,2));
    ver(i,2)=sum(disp(i*48-23:i*48,3));
end
hor(:,2)=-hor(:,2);
ver(:,2)=-ver(:,2);
%~~~~~~~~~~~~~~~~~~~~~~~~~~~~~~~~~~~~~~~~~~~~~~~~~~~~~~~~~~~~~~
% Calculating the difference between the curves
%~~~~~~~~~~~~~~~~~~~~~~~~~~~~~~~~~~~~~~~~~~~~~~~~~~~~~~~~~~~~~~
NOP=50;
minR=exphor(1,1);
minC=hor(1,1);
minABS=max(minR,minC);
maxR=max(exphor(:,1));
maxC=max(hor(:,1));
maxABS=min(maxC,maxR);
intPOINTS=minABS:(maxABS-minABS)/NOP:maxABS;
calR=interp1(exphor(:,1),exphor(:,2),intPOINTS);
calC=interp1(hor(:,1),hor(:,2),intPOINTS);
for i=1:size(calR,2)
    e1(i,1)=(calR(i)-calC(i))/1000;
end
NOP=50;
minR=expver(1,1);
minC=ver(1,1);
minABS=max(minR,minC);
maxR=max(expver(:,1));
maxC=max(hor(:,1));
maxABS=min(maxC,maxR);
intPOINTS=minABS:(maxABS-minABS)/NOP:maxABS;
calR=interp1(expver(:,1),expver(:,2),intPOINTS);
calC=interp1(ver(:,1),ver(:,2),intPOINTS);
for i=1:size(calR,2)
    e2(i,1)=(calR(i)-calC(i))/1000;
end
e=[e1;e2];
**************************************************************
```

Input to the above MATLAB function is a vector of parameters **x**. The vector collects four entries representing normalized values of sought parameters. The first lines of the code are used to transform normalized values into those that will be used in input files. Note that Young's modulus doesn't belong to the sought parameters, and it's kept fixed and considered as known parameter, so its value is given within the code as it is needed to compute stress–strain curve.

The routine further needs the existence of two external files with experimental results. In this case they represent two curves written in two separate files: one for the horizontal direction and another one for the vertical direction.

Further step is to transform the input file. For this case study, as previously mentioned, it is necessary to compute stress–strain curve defined by Eq. 4.4 in the tabular form, as a function of exponent of hardening (one of the sought parameters), and yield limit for the direction 1, as another sought parameter. This curve obviously depends also on Young's modulus that is unchangeable as previously mentioned. Since exponential hardening defined by Eq. 4.4 involves larger gradient of changes in first phase right after the yield limit, the tabular computation is divided in two steps: the first one is valid for the equivalent plastic strains between 0 and 0.01, where denser grid is used (i.e. with step 0.001). For larger values of equivalent plastic strains, the stress gradient drops down (it is the case for moderate values of hardening exponent) and therefore, in order to avoid large number of tabular values less dense grid is used for this range (i.e. with the step of 0.02).

Second part of the input file that we need to change considers the ratios (4.3). In particular, we need to compute only R_2 and R_4, since the first one is kept fixed and equal to 1, having therefore defined the yield limit for the first direction by a parameter σ_0. As already mentioned Hill's yield function definition in ABAQUS needs the values for all six ratios, so they are written into the new input file with values 1.

After the input file is changed and written on hard disk, the numerical simulation is performed by calling the ABAQUS solver. The numerical model is organized in such a way that the results are written into the external ASCII file. After the analysis is finished function `readfil2` is called in order to load results into MATLAB surrounding. The argument of the function is the name of ASCII file with the results, and its outputs are one vector and one matrix as discussed in Sect. 4.2.2. These results are further transformed into required force-displacement curves stored in matrices `hor` and `ver`. First column of the two matrices represents the displacements that are simply obtained by multiplying increment with prescribed displacement in the analyses. Second column represents the corresponding reaction forces. These are computed by a summation of all the components corresponding to one increment. In this case, both node-sets have 24 nodes which means that first 48 lines of the resulting matrix `disp` represent values of the reaction forces after the first increment, second 48 after the second one and so on. Each line has the reaction forces in both directions. However, considering the constrains, nodes belonging to node-set "left" will have only first component different from zero, since the other one is free to move, while those belonging to node-set "bottom" will have the second one different from zero. Therefore, the reaction force for the horizontal curve corresponding to i^{th} increment represents a summation of the members of matrix `disp` placed in the second column with the following indexes:

```
disp(i*48-47:i*48-24,2)
```

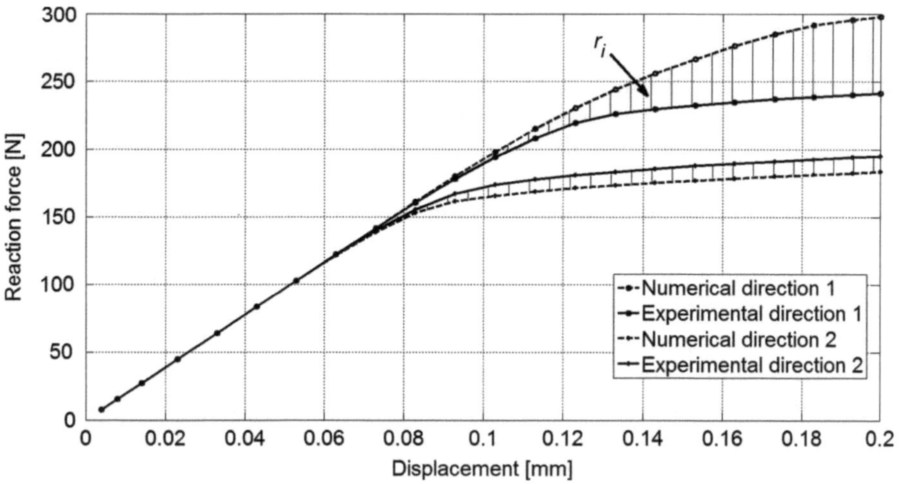

Fig. 4.9 Residuals for discrepancy function that quantifies the difference between two pairs of curves

while the one for the vertical curve is given by a summation of the third column of the members with the following indexes:

```
disp(i*48-23:i*48,3)
```

The final part of the routine quantifies the differences between the two curves and builds the vector of residuals. In this case the response to the experiment consists of two force-displacement curves that are confronted to their computed counterpart. The residuals are computed on the same way as in previous case, namely they represent the difference in the reaction forces computed for the number of grid points fixed over displacement range. This is illustrated in Fig. 4.9. The number of grid points is defined within the program by the variable `NOP`.

In order to perform the testing of the whole process, also for this case study, we will need pseudo-experimental curves that are created for some given set of parameters. For the exercise here we will create numerical results for the following values of the parameters:$\sigma_1 = 400\text{MPa}$, $\sigma_2 = 320\text{MPa}$, $\tau_{12} = 230\text{MPa}$ and $n = 0.1$. The resulting two curves are saved in two ASCII files: `exphor.txt` and `expver.txt`.

After this preparative work we can proceed to the final step by writing an optimization routine that will solve the resulting minimization problem. Considering that we know exactly the target values for the parameters we can perform the test of the overall accuracy of the inverse analyses procedure.

4.2.4 Solving the Optimization Problem

As a final part of inverse analysis procedure we need to write a routine that will solve resulting optimization problem. Also in this case we will make a use of Hessian approximation and avoid computing second derivatives. In this problem discrepancy function is less convex, which is usually the case when experimental data are more numerous (here we have two force-displacement curves). It means that the number of local minima, maxima or saddle points is elevated, and in these regions it is rather difficult to find a Newton direction computed by modified Hessian matrix that would lead to a reasonable function reduction.

The problem originates in the type of modification of the Hessian matrix that is used. When a true Hessian matrix is used, in the situations in which it is not positive-definite it can be modified by some of the procedures described in Chap. 3 in order to ensure its positive definitiveness and therefore yield the descending step. It is also important to know how much the Hessian differs from positive-definitiveness. In the two variable space, if for example one eigenvalue is positive and the other one is negative it indicates a saddle point, so still the direction that can be followed as it is minimizing the function. On the other hand, if both eigenvalues are negative it indicates maximum of the function, and modified Newton direction in this case doesn't have much sense, since the true one is pointing to a local maximum.

With the approach used for least squares, Hessian matrix is approximated by (2.7) that is a quadratic form, which means that the modified Hessian will be always positive-semi-definite. Using this modification instead of the real Hessian, valuable information on how much Hessian matrix defers from positive-definitiveness is lost. This information otherwise can be used as an indication whether to use modified Newton direction, or to adopt some other possibility (e.g. steepest descend). Therefore, in the algorithms that relay on modified Hessian matrix there is no sense in implementing a modification of Hessian in order to force it to be positive-definite (parts of the program that we saw in earlier implementation of different optimization algorithms given in this book). However, the direction computed with Hessian modified in this way (i.e. by Eq. 2.7) may be ineffective or may even not decrease the function.

In order to overcome previous problem a modified version of dog-leg algorithm is used. The procedure is the same as in a standard algorithm except that, in the case of potential step rejection (i.e. if the ratio (2.22) is negative for the dog-leg solution), before actual rejection the algorithm computes first the reduction of real objective function for the Cauchy point. If it reduces the function, the step is accepted and the algorithm proceeds to the next iteration. This scheme is presented in Fig. 4.10, where dark grey fields are representing computationally "expensive" parts of the procedure, as they involve numerical simulations.

Form the figure it is clear that this scheme, once that it faces previously discussed problem, performs an additional simulation in order to compute function reduction for Cauchy point. In case when also this step is not producing reduction of the function it may be argued that it's an additional waste of time for the step that should be rejected. However, this modification is inserted to tackle problems when Newton direction computed by modified Hessian matrix is not leading to the reduction of the function and so reducing of trust region will lead to further step rejections as the size of the region is not the cause of the problem. The logic of this

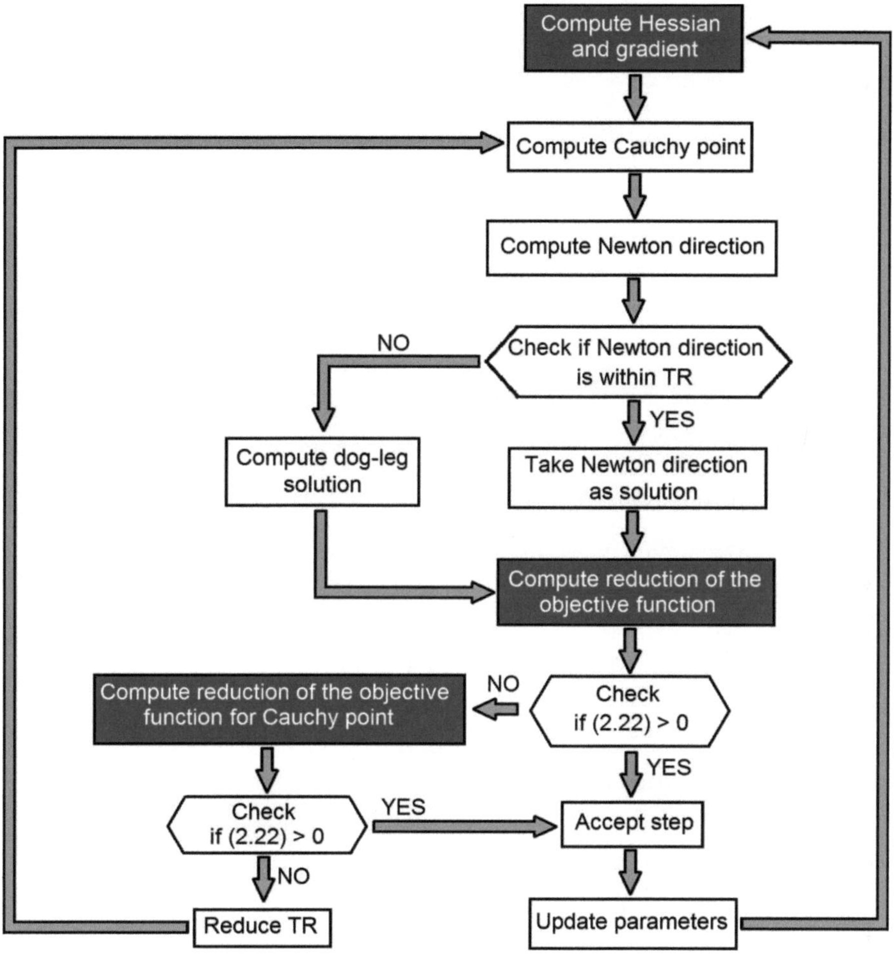

Fig. 4.10 Schematic representation of modified dog-leg algorithm

scheme is to use in these situations solution along steepest descent and to obtain some reduction of the function if not the best one. If this reduction is much smaller than what is predicted still, the trust region will be reduced for next iteration. With this approach, optimization algorithm avoids being trapped in the loop of continues reduction of trust region and repeatable step rejections when modified Newton direction is a misleading one. Instead it takes a Cauchy point solution and moves on in the optimization.

Implementation of this algorithm is given in the MATLAB listing below.

```
*************************************************************
% Trust region algorithm with modified dog-leg approach for
% sub-problem
clear
clc
%~~~~~~~~~~~~~~~~~~~~~~~~~~~~~~~~~~~~~~~~~~~~~~~~~~~~~~~
% Setting the options
minchg=1e-8; % Minimum change in parameters
MAXIT=30; % Maximum allowed number of iterations
guess=[1.40;1.35;0.7;0.4];
pert=1e-4; % Perturbation for the first derivatives
res=10;
TRrad=0.2;
mingrad=1e-6; % Gradient value at which to terminate optim.
%~~~~~~~~~~~~~~~~~~~~~~~~~~~~~~~~~~~~~~~~~~~~~~~~~~~~~~~
% Computing for the first time value of function
eV=disfunfem(guess);
e0=0.5*eV'*eV
%~~~~~~~~~~~~~~~~~~~~~~~~~~~~~~~~~~~~~~~~~~~~~~~~~~~~~~~
% Optimization cycle
iter=0;
while res>1e-6
itiner(iter+1,1:4)=guess';
itiner(iter+1,5)=e0;
iter=iter+1;
if iter==60
    pert=5e-5;
end
[HESS,grad]=comhessapp(@disfunfem,guess,pert,eV);
% Checking the gradient
if grad'*grad<mingrad
    break
end
stpdsc=-grad/norm(grad);
newton=-inv(HESS)*grad;
accepted=0;
rejected=0;
%~~~~~~~~~~~~~~~~~~~~~~~~~~~~~~~~~~~~~~~~~~~~~~~~~~~~~~~
% Computing step
while accepted<1
if norm(newton)<TRrad
    pDL=newton; % Dog Leg step
else
    % Finding the Cauchy point
    pc=cauchypnt(e0,stpdsc,grad,HESS,TRrad);
    % Finding minimizer within trust region (Dog Leg step)
    diff=newton-pc;
    dimV=size(newton,1);
    cf=[0,0,-TRrad^2];
```

```
    for ii=1:dimV
        cf(1)=cf(1)+diff(ii)^2;
        cf(2)=cf(2)+2*pc(ii)*diff(ii);
        cf(3)=cf(3)+pc(ii)^2;
    end
    alfa=max(roots(cf)); % Taking the positive root
    pDL=pc+alfa*diff;
end
    predred=-(pDL'*grad+0.5*pDL'*HESS*pDL);
    guess1=guess+pDL; % Next iterate
    eV=disfunfem(guess1);
    e1tr=0.5*eV'*eV;
    realred=e0-e1tr;
    ratio=realred/predred;
    if ratio<0
        TRrad=TRrad/1.2;
        disp('Step is rejected but trying Cauchy point!!!')
        pc=cauchypnt(e0,stpdsc,grad,HESS,TRrad);
        guess1=guess+pc;
        eV=disfunfem(guess1);
        e1tr=0.5*eV'*eV;
        realred=e0-e1tr;
        ratio=realred/predred;
        if ratio<0
            rejected=rejected+1;
            disp('Step is rejected!!!')
        else
            accepted=1;
        end
    else
        accepted=1;
        if ratio<0.2
            TRrad=TRrad/1.2;
        end
        if ratio>0.75
            TRrad=TRrad*1.2;
        end
    end
    if rejected==6
        accepted=1;
        guess1=guess; % Since there was no improvement
    end
end
itiner(iter+1,1:4)=guess1';
itiner(iter+1,5)=e1tr;
e0=e1tr;
guess=guess1;
res=e1tr;
if iter>MAXIT % Terminated after MAXIT iteration
```

```
    res=0;
end
% Checking convergence options
if abs(itiner(iter+1,1)-itiner(iter,1))<minchg ||
abs(itiner(iter+1,2)-itiner(iter,2))<minchg;
    res=0;
end
end
************************************************************
```

This main routine uses also previously listed MATLAB function that computes the value of discrepancy function together with routines for Hessian approximation (`comhessapp`) and computation of Cauchy point (`cauchypnt`). The rest of the routine is organized in the similar way as other already given programs. The changeable optimization portions are set at the beginning while the result is stored in the `itiner` matrix in the same manner as in previous case study.

4.2.5 Results of Inverse Analyses

With the last written code the inverse problem is fully prepared. Phase of testing the whole procedure proceeds using previously created pseudo-experimental data.

The inverse problem should be solved starting from different initialization points. Optimization parameters are set as follows: initial Trust Region is set to 0.2; perturbation for derivates are set to 1E-4, the optimization should be terminated when gradient is smaller than 1E-5, when minimum change in parameters between two iterations is less than 1E-8 (for normalized parameters) or when the number of iterations exceeds 30.

The first initialization started from the following parameter values: $\sigma_1 = 560\text{MPa}, \sigma_1 = 432\text{MPa}$, $\tau_{12} = 161\text{MPa}$and $n = 0.04$. The optimization was terminated after nine iterations since the changes in parameter values from eighth to ninth iteration was less than what was specified by tolerance. Largest error was on the assessed yield limit τ_{12}, which was evaluated as 234 instead of 230, that corresponds to less than 2% of difference. Graphs in Fig. 4.11 are visualizing the monotone drop of the objective function together with normalized parameter values within the iterations.

Trust region radius at the beginning of the optimization was 0.2 and at the end it was enlarged to 0.228. There were totally two step rejections, where resulting step was actually increasing the objective function: in the iteration 7 and 8. In both cases Cauchy point solution computed afterwards was obtaining the reduced function value and was therefore accepted as a solution.

An optimization problem should be solved couple of times starting from different initialization values for parameters in order to verify that the same solution is obtained independently from the starting point. As a usual practice, converged set

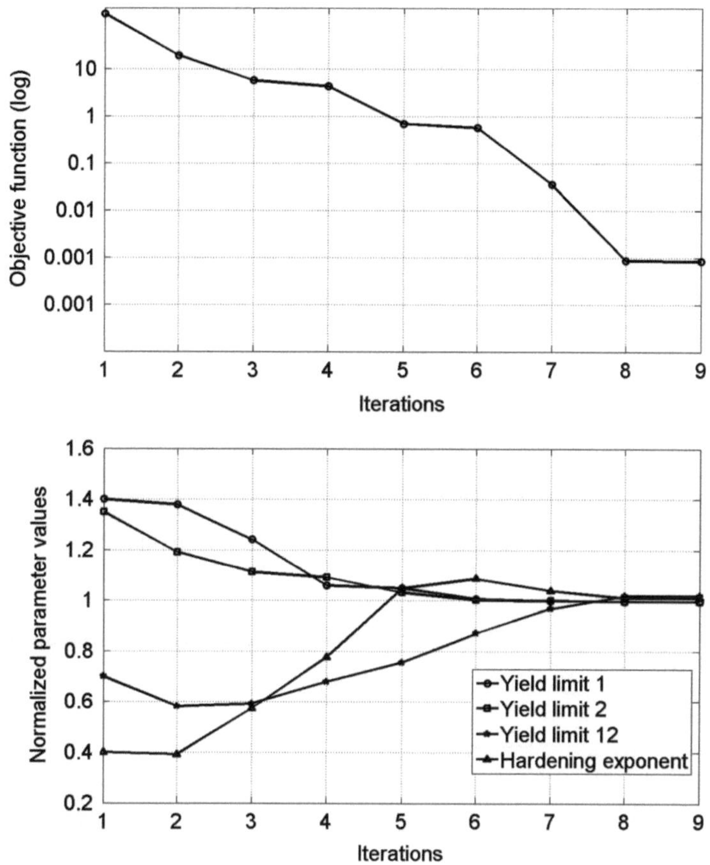

Fig. 4.11 Result of optimization: initialization 1 – reduction of the objective function (*up*) and changes of normalized parameters through iterations (*down*)

of parameters can be treated as a solution of to problem if it is repeated for at least three different initializations.

Inverse problem here considered is solved another time, starting from different initial parameter values with respect to those used in the first optimization. In the second optimization parameters had the following starting values: $\sigma_1 = 320$MPa, $\sigma_1 = 512$MPa, $\tau_{12} = 276$MPa and $n = 0.06$. Also in this case initial values were quit far from the target – more than 50% difference as an average. Tolerances were set to the same values as in the previous case, and the optimization was terminated after ninth iteration, converging to the following parameter set: $\sigma_1 = 399$MPa, $\sigma_1 = 319.3$MPa, $\tau_{12} = 233.3$MPa and $n = 0.1005$ Fig. 4.12.

The example treated in this Case study proved to be well posed, as the solution is not dependant on the initialization point. This can be also checked for some other set of parameters used as targets, which is usually the practice in full verification of the inverse problem. This part is left to the reader as an additional exercise.

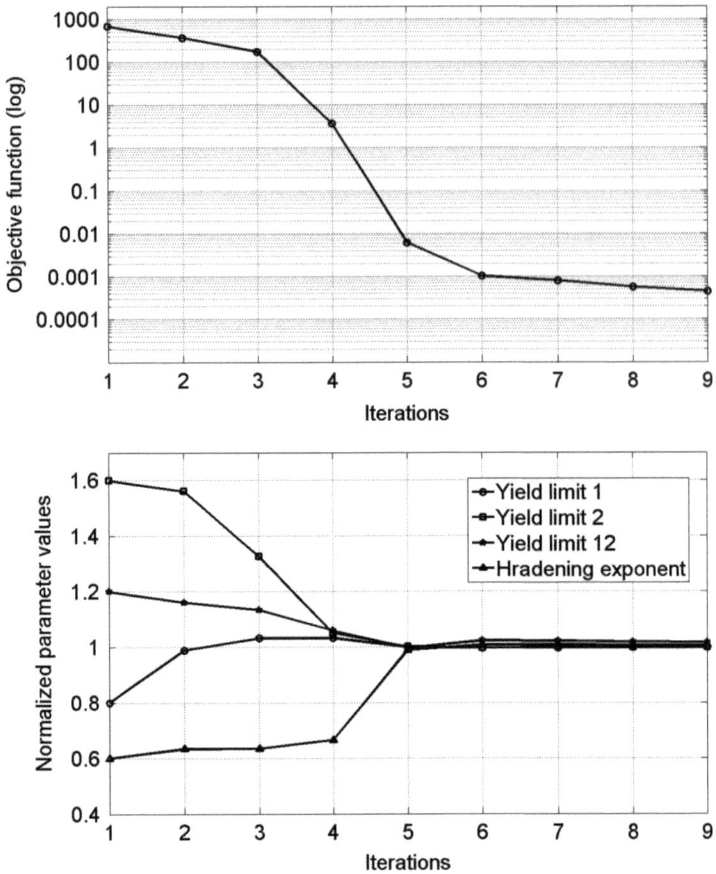

Fig. 4.12 Result of optimization: initialization 2 – reduction of the objective function (*up*) and changes of normalized parameters through iterations (*down*)

4.3 Summary

In this chapter we saw how to design a fully working Inverse Analyses procedure in order to assess some needed parameters. The problem was solved here by writing a set of MATLAB programs that are eventually put together into one fully automatic procedure.

Within two different case studies treated here, we saw all the steps needed to be put together in order to build required procedure. The first step was to select experiment and some measurable quantities that will be taken as a quantification of the response to the experiment. These should be selected in a way so that measurable quantities are sensitive to the changes in sought parameters. In more complicated identification procedures, when it is not that trivial to select measurable quantities and the experiment, it is useful to perform a sensitivity analyses in order to quantify the sensitivity of measured quantities with respect to sought parameters, and, if necessary to change experiment and/or measurable quantities. In the two examples studied here, it was quit intuitive to assume that selected measurable quantities are sensitive to sought parameters. This assumption proved to be good since in both procedures all of the parameters were well identified.

Second step consisted in building of numerical model that will simulate the experiment. In the two examples treated in this chapter, a commercial code ABAQUS was used for the purpose. This code has, for the present purpose a very suitable feature, that the model can be completely defined (and changed) by a human readable ASCII file. This feature makes the interaction with MATLAB easier. An input file that defines numerical model was divided in couple of parts, in a way that changeable part of the model is isolated within a single (and simple) text file.

Third step consists of writing a MATLAB function that will quantify discrepancy between experimental and computed data. Within this function experimental data are loaded from external file, an input file for ABAQUS model is loaded and required parameters are changed, the FE analysis is performed, and resulting data are loaded in MATLAB in order to be compared with experimental ones. For this purpose, another useful feature of ABAQUS is exploited, namely the needed results are written within an ASCII file so that they can be easily loaded in MATLAB surrounding. An additional MATLAB function was written and listed above that can be used to extract needed data from resulting ASCII file.

Fourth part consisted in writing the optimization algorithm that is used to solve the resulting minimization problem. For this purpose, algorithms presented in Chap. 2 are adopted, and their full listings are given in the chapter.

Finally, the whole procedure is tested by the use of pseudo-experimental data. This data, as we saw in the chapter, represent a computer generated responses to the experiment, which are treated as truly experimental data within inverse analysis procedure. The advantage of their use is that we know exactly the solution of the problem as it should match the parameters previously used to generate pseudo-experimental data. It is therefore easy to evaluate the overall accuracy of the procedure.

Further test, which are usually performed as part of the verification should consider influence of the measuring noise. With the checks performed in the case studies presented in this chapter we could see that the procedure is well posed and that selected measurable quantities are enough for accurate estimates of sought parameters. Since with the real experiment there will be always some measuring error it is important to understand to which extent it influences the error on assessed parameters. A few words on this topic will be given in the following chapter.

This chapter presented what is nowadays called traditional approach to inverse problems when test simulations are performed by FE models. However an inverse problem based on FE simulations can be sometimes time consuming as the simulations are performed many times. A modern approach to inverse analysis will be discussed in the subsequent chapter, where soft computing methods presented in Chap. 3 will be employed for the present purpose. It is important to mention that, traditional approach given in this chapter still represents an important part, as it is anyhow used at least in some phases of the design of inverse analyses procedures.

References

1. HKS Inc, Pawtucket, RI, USA. ABAQUS/Standard, Theory and User's Manuals, release 6.2-1 (1998)
2. Lubliner, J.: Plasticity Theory. Dover, Mineola (2008)
3. Crisfield, M.A.: Non-linear Finite Element Analyses of Solids and Structures. Wiley, Chichester (1997)
4. Green, D.E., Neale, K.W., MacEwen, S.R., Makinde, A., Perrin, R.: Experimental investigation of the biaxial behavior of an aluminum sheet. Int. J. Plast. **20**, 1677–1706 (2004)
5. Hannon, A., Tiernan, P.: A review of planar biaxial tensile test systems for sheet metal. J. Mater. Process. Technol. **198**, 1–13 (2008)
6. Hild, F., Roux, S.: Full Field Measurements and Identification in Solid Mechanics Volume 1. KMM-NoE, Warsaw (2007)

Chapter 5
Modern Approach to Inverse Analyses

By surveying the scientific literature, it can be observed that inverse problems are nowadays becoming more and more popular. Today, there are couple of international journals with high impact factors devoted to this class of problems. Apart of them there are many recently published books that are dealing with mathematical programming in the context of inverse analyses, several are cited within bibliography in previous chapters of this book.

The field of application of this fast growing scientific branch is vast. Some of them are listed in Chap. 1, but in general, this theory can have a successful application in any problem in which the knowledge of parameters, boundary conditions, initial conditions, or commonly speaking some not measurable information of the system are needed, while the results (or effects) are known.

An important group of inverse problems are parameter identification problems in structural engineering. Today there are many successful examples of applying this theory in material characterization and structural diagnosis. In engineering practice quite frequently occurs a need to characterize materials of a working components in order to assess the level of accumulated damage. In these satiations it is important that the test is non-destructive and that it can be applied on a working component in-situ.

Combining for example instrumented indentation test, mentioned in Chap. 1, with inverse analyses theory, material characterization can be performed on more sophisticated way with respect to semi empirical approaches (i.e. using formulae to compute reduced Young's modulus based on the indentation curve). The need however, to solve resulting inverse problem fast and in situ (i.e. on a "small" computer) makes the approach challenging from the computational point of view. Obviously for this class of problems a traditional approach based on finite element simulations is not an adequate one.

The use of FEM in the contexts of inverse analysis for structural problems is suitable for laboratory based problems as they involve time consuming analysis and, depending on a problem can even last for days. When there is a need to solve inverse problem fast and in-situ, alternative strategies should be used. This chapter will describe one possible approach, which uses POD-RBF algorithm presented in Chap. 3, in this context.

V. Buljak, *Inverse Analyses with Model Reduction*, Computational Fluid and Solid Mechanics, DOI 10.1007/978-3-642-22703-5_5,

5.1 On-Line Off-Line Approach

Examples treated in previous chapter showed that the main "bottle-neck" in terms of computing time involved for solving the inverse problem is a need to perform finite element simulations. The rest of the computations are fast and are practically done immediately, in no time at all. The iterative nature of the optimization algorithms requires recurrent simulations, with total number which can reach order of magnitude of 100, when number of sought parameters is moderate (e.g., four or more). However numerical models used in these simulations are quite similar one to another, with only couple of parameters varying (namely sought parameters). This fact suggests that in this field quite effectively some model reduction techniques can be applied, in order to make the problem much faster and more robust.

Today model reduction techniques represent quite popular research subject. There are many successful applications of POD used in so-called *a priori* model reduction procedures (see e.g., [1]). Interesting examples of application of such methodology can be found in simulations of real-time deformable models of non-linear tissues [2]. Building of reliable deformable numerical models of soft tissues represents a challenging task in computational mechanics. It has a wide applicability in design of haptic simulators used in virtual surgery. The physical phenomena simulated in these problems are extremely non-linear (namely, non linear visco-elastic material behavior combined with large deformations and large displacements) and as such represent a computationally demanding task that practically cannot be tackled by traditional FEM approaches if it should be done in a real-time. Similar methods based on Proper Generalized Decomposition (PGD) are developed and applied in computational fluid dynamics and in computational fracture mechanics (see e.g., [3, 4]).

Most of these techniques have the two computational phases: off-line and on-line phase. In the off-line phase some relatively heavy computation are performed but usually they are done once-for-all. Later, in on-line phase, previous results are used in a smart way, and some lighter computations are performed in order to have the results fast, practically in a real-time.

Applying this off-line on-line approach to the inverse problems similar strategy can be developed. Considering that within parameter identification procedures, as previously mentioned, simulations that need to be performed are very similar one to another, as they differ only by a few parameters, it is possible to think of an approach in which within a first phase (off-line phase) a set of these simulations should be performed, in order to use obtained results in s latter, on-line phase.

This approach is essentially the same as the one when Artificial Neural Networks are applied to solve inverse problem. Without entering into the detailed description of neural networks, which can be found in some other excellent books (see e.g., [5, 6]), let us just say couple of words about main principles of their work. Artificial Neural Networks (ANN) represent an interpolation tool which can be used for the prediction of system responses. Calibration of ANN, in jargon called "training", is the phase in which coefficients of this interpolation are determined (called weights and biases). For the training phase it is necessary to have a certain number of input–output pairs, called "patterns". After the training phase is performed network has a so-called generalization feature which means that it can provide outputs also for those inputs that were not present within the training patterns.

In order to apply ANN to solve inverse problems in parameter characterization it is necessary to produce first a required number of training patterns. It practically means that we need first to perform a set of FE simulations in which the sought parameters will be varied within some reasonable range, usually selected by the nature of the problem for which the ANN is designed. After these "heavy" computations are executed, the network is ready to be used for any further identification procedure. The training of the network is time consuming but it is done once-for-all and represents an off-line phase. Later on, identification procedure that is done on-line, involves only a simple matrix multiplication and it is therefore computationally "light". Examples of the application of ANN in parameter identification for structural problems can be found in [7] and [8].

Alternative strategy of using the results previously generated by FE simulations, could be by calibrating a POD-RBF model which can be later used in on-line phase within iterative optimization algorithms.

5.2 The Use of Pod-RBF Within Inverse Analysis Context

Numerous examples of parameter identification in structural context that can be found in scientific literature evidence that inverse analysis can be used rather successfully in this context. However, as mentioned above, they are not applicable for fast computations due to the extended time required by FE simulations. A logical path of overcoming this problem is to replace the system response computations by some other faster technique and to preserve the rest of the structure.

From the examples treated in Chap. 3 it is quite clear that this task can be fulfilled with forward operator based on POD-RBF algorithm. This strategy guarantees all the robustness and applicability of traditional Inverse Analyses approach already evidenced in vast literature, provided that within the inverse analyses procedure, FE simulations are replaced with some other tool with the same accuracy. The scheme of this strategy is visualized in Fig. 5.1.

The advantage of present approach with respect to, for example ANN is that, for the same cost in terms of computing time needed for generating a set of snapshots by FE simulations, a more controllable identification procedure can be obtained. If the inverse problem is well designed in terms of optimization of experiment and measurable quantities, optimization algorithms discussed in previous chapters are guaranteeing the convergence to the solution (e.g., those based on Cauchy point since they have a global convergence property). On the other hand, examples shown in Chap. 3 proved that reduced model based on POD-RBF can have controllable

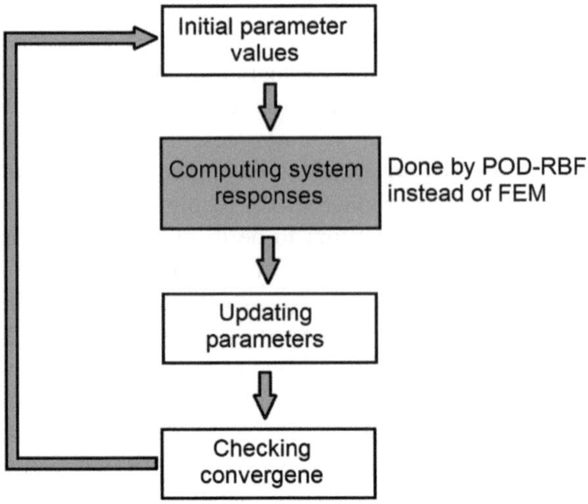

Fig. 5.1 Schematic representation of IA procedure based on POD-RBF

error and for the practical purposes the difference between system responses computed by it, and the one computed by FEM can be negligible. All this contributes to the conclusion that inverse analyses procedure based on this approach can be fast and robust with effective control of error on estimates.

Combining previous elements it is relatively easy to enclose all of the parts into a simple software, which can be trained for a given characterization problem, and further routinely used.

5.3 Example of the Use of Pod-RBF in Inverse Analyses

5.3.1 Design of Software for the Assessment of Parameters

We will show now how, an easy to use software, based on inverse analyses and POD-RBF model as forward operator, can be designed for characterization of anisotropic plastic parameters based on bi-axial test of thin plate (example treated in previous chapter).

At the beginning we will perform, as a part of off-line process, a generation of snapshots by varying parameters in order to cover some range of interest. For example we can generate total number of 875 snapshots over a regular grid of points in parameter space with the following limits: σ_1 from 300 to 460 MPa with step 40 MPa; σ_2 from 300 to 460 MPa with step 40 MPa; σ_{12} from 180 to 260 MPa with step 20 MPa. For this purpose a simple MATLAB code can be used that automatically updates the input parameters, runs analyses and stores resulting vector into a snapshot matrix. The listing of such MATLAB routine is given below

```
************************************************************
% Program that generates a snapshot matrix
clear
points=0.02:0.02:1;
numAN=0;
for sig1=300:40:460
 for sig2=300:40:460
   for sig12=180:20:260
     for N=0.05:0.025:0.2
      parY=sig1;
      parY2=sig2;
      parY12=sig12;
      parN=N;
      parE=200000;
      numAN=numAN+1;
      kpar(:,numAN)=[parY;parY2;parY12;parN];
% Changing the input file
%~~~~~~~~~~~~~~~~~~~~~~~~~~~~~~~~~~~~~~~~~~~~~~~~~~
      br=0;
      elstrain=parY/parE;
      for plstrain=0:0.001:0.01
        br=br+1;
        ssc(br,1)=parY^(1-
parN)*parE^parN*(elstrain+plstrain)^parN;
        ssc(br,2)=plstrain;
      end
      for plstrain=0.02:0.02:0.3
        br=br+1;
        ssc(br,1)=parY^(1-
parN)*parE^parN*(elstrain+plstrain)^parN;
        ssc(br,2)=plstrain;
      end
      R2=parY2/parY;
      R4=parY12/(parY/sqrt(3));
      f_fil=fopen('material.inp');
      s=fscanf(f_fil,'%c');
      fclose(f_fil);
      position=strfind(s,'Plastic');
      linebreak=s(position+7:position+8);
      begg=position+9;
      num=begg;
    newPL=[num2str(ssc(1,1)),',',num2str(ssc(1,2)),linebreak];
      for i=2:size(ssc,1)
       line=[num2str(ssc(i,1)),',',num2str(ssc(i,2))];
       newPL=[newPL,line,linebreak];
      end
     newPOT=['*Potential',linebreak,'1,
',num2str(R2),',1,',num2str(R4),',1,1'];
     s1=[s(1:begg-1),newPL,newPOT];
     f_fil=fopen('material.inp','w');
     fprintf(f_fil,s1);
```

```
      fclose(f_fil);
% ABAQUS run
%~~~~~~~~~~~~~~~~~~~~~~~~~~~~~~~~~~~~~~~~~~~~~~~~~~
! abaqus j=bi_ax_inp interactive
% Reading .fil file
%~~~~~~~~~~~~~~~~~~~~~~~~~~~~~~~~~~~~~~~~~~~~~~~~~~
       [disp,indcur]=readfil1('bi_ax_inp.fil');
       NINC=size(indcur);
       for i=1:NINC
         hor(i,1)=indcur(i);
         ver(i,1)=indcur(i);
         hor(i,2)=sum(disp(i*48-47:i*48-24,2));
         ver(i,2)=sum(disp(i*48-23:i*48,3));
       end
         hor(:,2)=-hor(:,2);
         ver(:,2)=-ver(:,2);
         snapHOR=interp1(hor(:,1),hor(:,2),points);
         snapVER=interp1(ver(:,1),ver(:,2),points);
         snapH(:,numAN)=snapHOR';
         snapV(:,numAN)=snapVER';
       end
     end
   end
end
************************************************************
```

The structuring of the file is very similar to the one already presented in Chap. 4. Some of the things needed to be performed are exactly the same like there, namely changing of input file, running analysis and reading results. The rest of the routine consist of placing the resulting vector into a snapshot matrix.

After this off-line computing phase is finished, it is possible to create POD-RBF model that gives as output force-displacement curves. In this example, we have calibrated two different models: One for the horizontal curve and second one for the vertical curve. In order to fully define a curve it is needed to have a set of pairs of displacement and corresponding reaction forces. However, as all the simulations are done with the same prescribed maximal displacement, it is therefore possible to have a fixed grid of points over displacement axis, and to take corresponding forces as measured quantities that will be entries of snapshots. In this exercise, a grid of 50 points is adopted for the displacements between 2% and 100% of maximum applied displacement with the step of 2%. As far as the RBF type is concern, also in this example inverse multiquadric one is used (Eq. 3.93 with $r = 1$.) Resulting two POD bases are both truncated after fifth direction with the ratio of the summation of all the neglected eigenvalues, and all of them, less than 10^{-6}. This calibration process therefore resulted in four matrices: $\mathbf{B}_1$, $\mathbf{B}_2$, Φ_1 and Φ_2. Any further system response is obtained by performing two matrix multiplication using (3.92) – one for the horizontal curve and another one for vertical curve.

Using MATLAB Graphical User Interface (GUI) it is possible to design an easy to use software that can be used to determine plastic parameters based on bi-axial test just by loading the two resulting curves from the test. Listing bellow serves for that purpose.

```
************************************************************
function varargout = biaxial(varargin)
% BIAXIAL M-file for biaxial.fig
% Begin initialization code - DO NOT EDIT
gui_Singleton = 1;
gui_State = struct('gui_Name',       mfilename, ...
                   'gui_Singleton',  gui_Singleton, ...
                   'gui_OpeningFcn', @biaxial_OpeningFcn, ...
                   'gui_OutputFcn',  @biaxial_OutputFcn, ...
                   'gui_LayoutFcn',  [] , ...
                   'gui_Callback',   []);
if nargin && ischar(varargin{1})
    gui_State.gui_Callback = str2func(varargin{1});
end
if nargout
    [varargout{1:nargout}] = gui_mainfcn(gui_State,
varargin{:});
else
    gui_mainfcn(gui_State, varargin{:});
end
% End initialization code - DO NOT EDIT

% Executes just before biaxial is made visible.
function biaxial_OpeningFcn(hObject, eventdata, handles,
varargin)
handles.output = hObject;
% Update handles structure
guidata(hObject, handles);

% Outputs from this function are returned to the command line.
function varargout = biaxial_OutputFcn(hObject, eventdata,
handles)
varargout{1} = handles.output;

function edit1_Callback(hObject, eventdata, handles)
fileH=get(hObject,'String');
curH=char(fileH);
curhor=load(curH);
axes(handles.axes1);
plot(curhor(:,1),curhor(:,2),'LineWidth',2);
grid on
title('Force-displacement curve for horizontal direction')

% Executes during object creation, after setting all
properties.
function edit1_CreateFcn(hObject, eventdata, handles)
if ispc && isequal(get(hObject,'BackgroundColor'),
get(0,'defaultUicontrolBackgroundColor'))
    set(hObject,'BackgroundColor','white');
end
```

```
function edit2_Callback(hObject, eventdata, handles)
fileV=get(hObject,'String');
curV=char(fileV);
curver=load(curV);
axes(handles.axes2);
plot(curver(:,1),curver(:,2),'LineWidth',2);
grid on
title('Force-displacement curve for vertical direction')

% Executes during object creation, after setting all
% properties.
function edit2_CreateFcn(hObject, eventdata, handles)
if ispc && isequal(get(hObject,'BackgroundColor'),
get(0,'defaultUicontrolBackgroundColor'))
    set(hObject,'BackgroundColor','white');
end

% Executes on button press in pushbutton1.
% function pushbutton1_Callback(hObject, eventdata, handles)
% Optimizaiton by DOG-LEG Trust Region
clear
global fh bh fv bv kpar
load fh.txt
load bh.txt
load fv.txt
load bv.txt
load kpar.txt
% Setting the options
minchg=1e-8;
MAXIT=30;
guess=[1.40;1.35;0.7;0.4];
guess=rand(4,1);
pert=1e-4;
res=10;
TRrad=0.2;
eV=disfunpod(guess);
e0=0.5*eV'*eV;
% Optimization cycle
iter=0;
while res>1e-6
itiner(iter+1,1:4)=guess';
itiner(iter+1,5)=e0;
iter=iter+1;
if iter==60
    pert=5e-5;
end
[HESS,grad]=comhessapp(@disfunpod,guess,pert,eV);
stpdsc=-grad/norm(grad);
newton=-inv(HESS)*grad;
accepted=0;
```

```
rejected=0;
while accepted<1
if norm(newton)<TRrad
    pDL=newton;
else
    pc=cauchypnt(e0,stpdsc,grad,HESS,TRrad);
    diff=newton-pc;
    dimV=size(newton,1);
    cf=[0,0,-TRrad^2];
    for ii=1:dimV
        cf(1)=cf(1)+diff(ii)^2;
        cf(2)=cf(2)+2*pc(ii)*diff(ii);
        cf(3)=cf(3)+pc(ii)^2;
    end
    alfa=max(roots(cf));
    pDL=pc+alfa*diff;
end
    predred=-(pDL'*grad+0.5*pDL'*HESS*pDL);
    guess1=guess+pDL;
    eV=disfunpod(guess1);
    e1tr=0.5*eV'*eV;
    realred=e0-e1tr;
    ratio=realred/predred;
    if ratio<0
        TRrad=TRrad/1.2;
        pc=cauchypnt(e0,stpdsc,grad,HESS,TRrad);
        guess1=guess+pc;
        eV=disfunpod(guess1);
        e1tr=0.5*eV'*eV;
        realred=e0-e1tr;
        ratio=realred/predred;
        if ratio<0
            rejected=rejected+1;
        else
            accepted=1;
        end
    else
        accepted=1;
        if ratio<0.2
            TRrad=TRrad/1.2;
        end
        if ratio>0.75
            TRrad=TRrad*1.2;
        end
    end
    if rejected==6
        accepted=1;
        guess1=guess;
    end
end
itiner(iter+1,1:4)=guess1';
```

```
itiner(iter+1,5)=e1tr;
e0=e1tr;
guess=guess1;
res=e1tr;
if iter>MAXIT
    res=0;
end
% Checking convergence options
if abs(itiner(iter+1,1)-itiner(iter,1))<minchg ||
abs(itiner(iter+1,2)-itiner(iter,2))<minchg;
    res=0;
end
end
save itiner.txt -ascii -double itiner

% Executes on selection change in popupmenu1.
% function popupmenu1_Callback(hObject, eventdata, handles)
contents={get(hObject,'Value')};
global kpar
load itiner.txt;
itiner(:,1)=itiner(:,1)*(max(kpar(1,:))-
min(kpar(1,:)))+min(kpar(1,:));
itiner(:,2)=itiner(:,2)*(max(kpar(2,:))-
min(kpar(2,:)))+min(kpar(2,:));
itiner(:,3)=itiner(:,3)*(max(kpar(3,:))-
min(kpar(3,:)))+min(kpar(3,:));
itiner(:,4)=itiner(:,4)*(max(kpar(4,:))-
min(kpar(4,:)))+min(kpar(4,:));
selected=cell2mat(contents);
ITER=size(itiner,1);
if selected==1
    axes(handles.axes3);
    plot(1:ITER,itiner(:,1),'LineWidth',2);
    grid on
    title('Force-displacement curve for horizontal direction')
end
if selected==2
    axes(handles.axes3);
    plot(1:ITER,itiner(:,2),'LineWidth',2);
    grid on
    title('Force-displacement curve for horizontal direction')
end
if selected==3
    axes(handles.axes3);
    plot(1:ITER,itiner(:,3),'LineWidth',2);
    grid on
    title('Force-displacement curve for horizontal direction')
end
if selected==4
    axes(handles.axes3);
    plot(1:ITER,itiner(:,4),'LineWidth',2);
```

```
    grid on
    title('Force-displacement curve for horizontal direction')
end

% Executes during object creation, after setting all
% properties.
function popupmenu1_CreateFcn(hObject, eventdata, handles)
if ispc && isequal(get(hObject,'BackgroundColor'),
get(0,'defaultUicontrolBackgroundColor'))
    set(hObject,'BackgroundColor','white');
end
*****************************************************************
```

As addition to previous listing also a MATLA fig file with the same name is needed in which the outline of the developed GUI will be designed. Here we are not going to describe the way GUI is designed in MATLAB and readers should refer to MATLAB help for this issue [9].

The outline of designed GUI is visualized in Fig. 5.2. Above listed MATLAB routine programs the behavior of all existing elements in it. Therefore, in the code we have two functions connected with two text input fields that are used to supply the file names where horizontal and vertical curves are stored. Further, within these functions resulting curves are plotted on graphs 1 and 2 that are placed on the left side of the window.

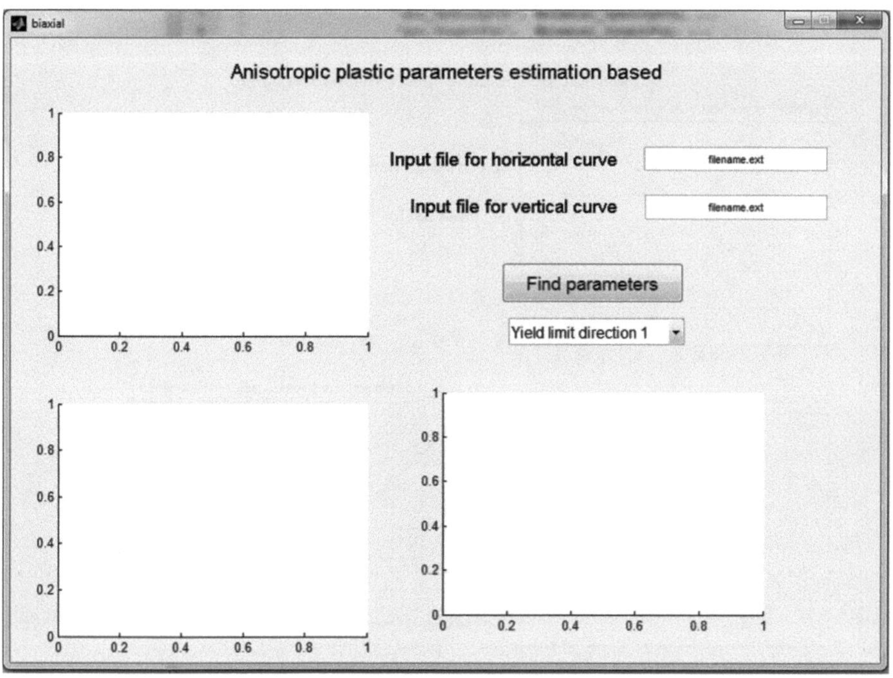

Fig. 5.2 The outline of GUI for plastic parameter estimation

After the input files are inserted and graphs are visualized program has everything that is needed to perform optimization in order to find four plastic parameters that are minimizing the difference between the experimental curves and those numerically computed. For this purpose a very similar function to the one already listed in Chap. 4 is associated to push-button "Find Parameters". The main difference in this function with respect to the one given in Chap. 4 is that here the system response is computed by POD-RBF model, not by calling ABAQUS.

After clicking on "Find Parameters" button, the optimization problem is solved and resulting parameters can be visualized one by one by selecting them from the drop-down menu placed bellow "Find Parameters" button. For this purpose, part of the code is associated with the drop-down menu that identifies which of the parameters is selected, and plots its values in each iteration on the third graph placed in lower right corner of the window.

Initialization of the optimization process is done randomly, and therefore it is easy to repeat the process in order to verify that the assessed parameter values are independent on starting point. After the optimization the resulting window will look like it is visualized in Fig. 5.3.

Optimization problem solved was the same as in Chap. 4, with target values specified there, and graph 3 in this case visualizes the changes of the yield limit in direction 2. Figure shows that the value of this parameter converged to the target one practically after five iterations.

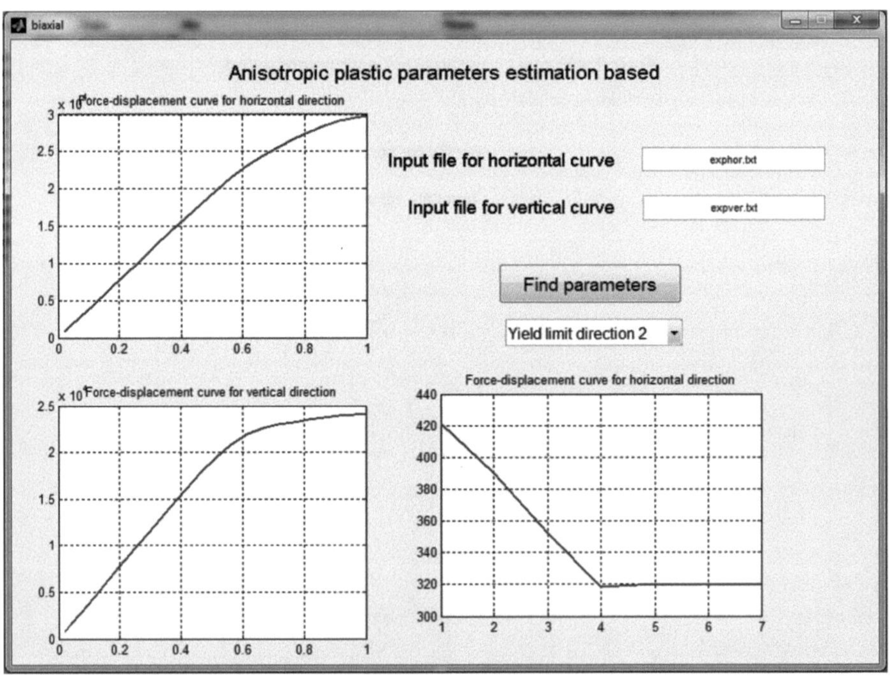

Fig. 5.3 Resulting window after the optimization is terminated

By having such a simple software the identification of parameters becomes easy, and immediate, since in this case one identification lasts for about 1 s. Furthermore, the identification procedure becomes a trivial matter since everything that one needs to do is to give the names of the two input files and to click on one button. All other work connected with computation of system responses and mathematical programming is enclosed inside this simple program and the user doesn't have to know anything about it. Of course, this program is calibrated for one type of experiment (in this case bi-axial test) and it requires some off-line computational work, but it is done once-for-all, and after that, an on-line phase is rather simple and can be routinely and repetitively used.

Applying this strategy to some other types of test (e.g. indentation test) it is rather easy to develop a simple software which can be used for fast assessment of parameters. Together with portable testing equipment it becomes a powerful tool for diagnostic analysis in situ on working components since the present approach in its on-line phase doesn't involve any heavy computation.

5.3.2 Detailed Pseudo-Experimental Testing of Inverse Analyses Procedure

With previous exercise we wanted to demonstrate how by the use of POD-RBF fast system response computation in the context of inverse analyses, the material characterization becomes an easy task. Having such a fast computing tool provides also some other advantages.

Already in Chap. 1 we were mentioning the need to test the identification procedure with pseudo-experimental data. This test should serve us to verify that the designed inverse analyses problem is well posed and that it can be successfully solved by selected optimization algorithm. In order to have a full control over the whole procedure, we are deliberately selecting what is called pseudo-experimental data, or data generated by a computer simulation for which we know exactly what is the solution in terms of sought parameters. However, even if the result of this preliminary test is positive, it may happen that, for some other parameter combination, we may encounter an ill-posed problem. This can happen with the indentation tests when only the indentation curve is used as experimental information. This problem is already emphasized earlier in the book when plastic parameters can be compensated and therefore the two different parameter sets can produce the same indentation curve. This compensation may be more evidenced for certain parameters and less for others. In other words by performing a test for only one parameter set, chosen as target, it is not guaranteed that the abovementioned compensation problem will be evidenced, and so the false conclusion of a well-posed inverse analyses problem can be derived. It is therefore very useful to perform a more detailed verification of the procedure to verify that it is well posed for different sets of parameters taken as target values.

More detailed investigation of the designed inverse analyses procedure are of course time consuming since it means that the problem needs to be solved for different pseudo-experimental results corresponding to different parameter sets. On the other hand, once that we have trained POD-RBF model for selected experiment it is relatively inexpensive to perform many different test in order to verify whether it's well posed in all zones of possible interest in the parameter space.

Let us focus on the problem of anisotropic plastic parameter characterization previously discussed. Once the snapshot matrix is computed we can compute matrices $\mathbf{B}_1$, $\mathbf{B}_2$, Φ_1 and Φ_2 that can be further used for fast computation of system responses. We can further test the accuracy of the whole procedure for different parameter sets as target values. In order to do that, we can generate another snapshot matrix, corresponding to parameter sets different from those used for the generation of POD basis. Resulting snapshots can be further used as pseudo-experimental data for the inverse analyses procedure. Assessed parameters are directly compared to their target values in order to see the overall error of the whole procedure.

Considering that the whole procedure is easily enclosed in the same programming surrounding (e.g., in MATLAB like in the examples treated here), the whole process can be fully automated. Algorithm for the possible solution of this testing is given in Fig. 5.4.

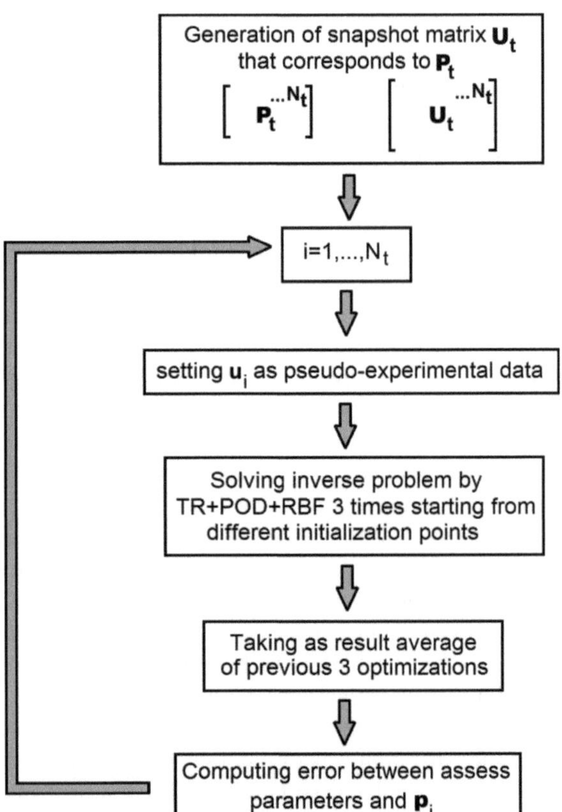

Fig. 5.4 Algorithm for testing of the whole TR + POR-RBF procedure

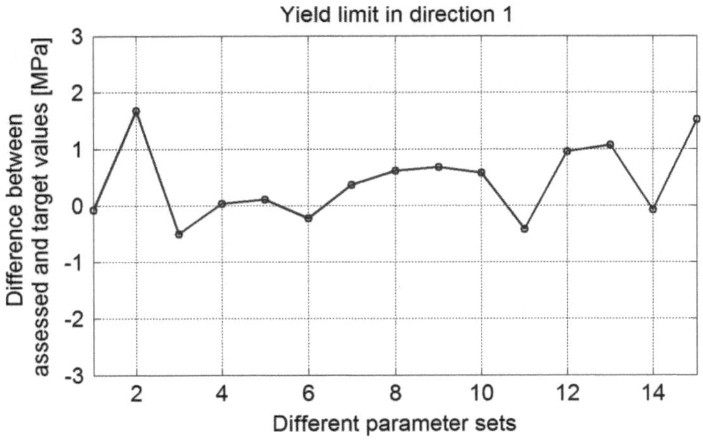

Fig. 5.5 Absolute error of assessed yield limit in direction 1 for 15 different parameter sets

Implementation of previous algorithm into a MATLAB code can be done by combining all previously given codes with slight modification. The listing of possible code is not given here and is left to the reader to be done as an exercise.

As a result of this test usually graphs of a type visualized in Fig. 5.5 are given for each parameter.

Figure 5.5 gives results obtained on a bi-axial test when the testing procedure is done on 15 different parameter sets. The graph shows that the maximum error on the assessed yield limit in direction 1 was less than 2 MPa. The result of this type is very good one, and if the similar one is obtained also for the other parameters, then it can be concluded that the inverse analyses procedure is well-posed for the whole tested range of parameters.

Considering computing times involved in such numerical test, we should note that the problem solved here involved 45 different inverse problems (15 different parameter sets and each inverse problem is solved three times starting from different initialization points). There were four parameters to assess, and by using a dog-leg trust region algorithm with Hessian approximation each iteration had five system response predictions. Finally taking as an average that each optimization had about 10 iterations, it gives a total number of system predictions equal to $45 \times 5 \times 10 = 2{,}250$. Performing such a test with traditional approach where system responses are computed by FEM, for this simple problem it would require about 10 h on an average computer. On the other hand using previously calibrated POD-RBF procedure the same result is obtained in a bit more than 5 s. It's important to emphasize that the calibration of reduced POD-RBF model included 875 simulations, a number that is smaller than what is otherwise required for even this small test if it should be performed with traditional approach, the fact that already for this simple case justifies a time consuming off-line phase. In more complicated examples, this type of test is more detailed and usually involves much more than only 15 material parameter sets. Furthermore, the same test is performed considering different levels of artificial noises introduced to the pseudo-experimental data in order to test the stability of the procedure. It is therefore clear that significantly more profound tests can be performed on the overall procedure once that the reduced model is calibrated.

At the end it is important to underline that the type of the test discussed here which results in graphs like the one visualized in Fig. 5.5, considers the overall accuracy of the procedure. The pseudo-experimental data in the exercise performed here are created by FE model, while the characterization procedure used POD-RBF reduced model for the system response computation. The fact that in all tested parameter sets the procedure converged practically to the same target values confirms the capability of the reduced model to give high accuracy results that are practically the same as what is computed by full FE numerical model.

5.4 Summary

This chapter gave an overview of a modern approach to the inverse analyses, where by concentrating some heavy computation in a priori phase (called off-line), fast and robust procedure can be designed to be executed routinely in on-line phase. The approach presented here combines reduced numerical model based on proper orthogonal decomposition and radial basis functions presented in Chap. 3, with iterative optimization algorithms. With such strategy, the core of the identification procedure is essentially the same as the one used in a traditional approach, except that the simulations required by optimization algorithm, are done with reduced model instead of FE one. This circumstance contributes to the huge acceleration of the whole procedure since, as we saw in Chap. 3, once that it is trained POD-RBF model computes the system response in a time shorter by about five orders of magnitude with respect to FEM.

This acceleration contributes to the overall robustness of the procedure, and in practical terms brings the methodology down to industrial level. The example treated in this chapter served to show that by implementing this strategy it is possible to build small stand-alone software which can be optimize for a given experiment and further routinely used in industrial practice.

Another big advantage of having such a fast computing tool, as we demonstrated in second example of this chapter, is the possibility to performe more profound tests of the whole inverse analysis procedure. When the inverse analyses procedure is designed in order to assess certain parameters, or in general some missing information, it is required to test its performance in many different rangers of parameters and for different levels of noises. We saw that this task usually involves the number of simulations that usually exceeds the number of simulations needed to train the POD-RBF model. This circumstance contributes to the justification of time consuming training of the procedure.

Finally, it should be underlined that, not by any means the purpose of this chapter was to show that FE simulations are obsolete since the same results can be obtained by much faster reduced models like POD-RBF. On the contrary, what we saw here and in Chap. 3 is that anyhow this reduced model needs a full numerical model for calibration and as a reference to check the accuracy level. It should be pretty clear by now that there is absolutely no sense in training a reduced model like the one presented here in order to use it further a few times since then there is no justification of the computing time spent for the training. On the contrary, in the situations in which it is needed to repeat many times simulations of numerical models that are very similar to each other and differ just by a few parameters it is reasonable to make a use of time savings provided by reduced numerical models. This is the case that we are facing in the inverse analysis in structural context and the goal of this chapter was to demonstrate which advantages can be gained when they are combined with POD-RBF procedure.

References

1. Ruckelynck, D., Chinesta, F., Cueto, E., Ammar, A.: On the a priori model reduction: overview and recent developments. Arch Comput Meth Eng **13**(1), 91–128 (2006)
2. Niroomandi, S., Alfaro, I., Cueto, E., Chinesta, F.: Real-time deformable models of non-linear tissues by model reduction techniques. Comput Meth Prog Biomed **91**, 223–231 (2008)
3. Rozza, G., Huynh, D.B.P., Patera, A.T.: Reduced basis approximation and a posteriori error estimation for affinely paramerized elliptic coefcive partial differential equations – application to transport and continuum mechanics. Arch Comput Meth Eng **15**(3), 229–275 (2008)
4. Nouy, A.: A priori model reduction through proper generalized decomposition. Comput. Meth. Appl. Mech. Eng. **199**, 1603–1626 (2010)
5. Hagan, M., Demut, H., Beale, M.: Neural Network Design. PWS Publishing, Boston (1996)
6. Waszczyszyn, Z.: Neural Networks in the Analysis and Design of Structures. Springer, Wien (1999)
7. Maier, G., Bolzon, G., Buljak, V., Miller, B.: Assessment of elastic-plastic material parameters comparatively by three procedures based on indentation test and inverse analyses. Accepted for publication in Inverse Probl. Sci. Eng. (January 2011)
8. Fedele, R., Maier, G., Miller, B.: Identification of elastic stiffness and local stresses in concrete dams by in situ test and neural networks. Struct. Infrastruct. Eng. **1**(3), 165–180 (2005)
9. The MathWorks Inc. Natick, USA. MATLAB 7 – Creating Graphical Users Interfaces (2007)

Index

V. Buljak, *Inverse Analyses with Model Reduction*, Computational Fluid and Solid Mechanics, DOI 10.1007/978-3-642-22703-5,

MIX
Papier aus verantwortungsvollen Quellen
Paper from responsible sources
FSC® C105338

If you have any concerns about our products,
you can contact us on
ProductSafety@springernature.com

In case Publisher is established outside the EU,
the EU authorized representative is:
Springer Nature Customer Service Center GmbH
Europaplatz 3, 69115 Heidelberg, Germany

Printed by Libri Plureos GmbH
in Hamburg, Germany